Why Are We Like This?

Zoe Kean lives in lutruwita/Tasmania, an island at the bottom of the world. An award-winning science writer with a focus on evolution, ecology, and the environment, she has published in *The Guardian*, the ABC online, *The Best Australian Science Writing 2022, 2023*, and *2024*, *Cosmos* magazine and with the BBC. Zoe honed her writing and radio skills with the best at the ABC Science Unit as the Darren Osborne Regional Science Cadet (2019). Her love for science cannot be contained to the page. Zoe also gives talks, makes science TikTok videos and regularly appears on live radio.

<www.zoekean.com/>

Why Are We Like This?

An evolutionary search for answers to life's big questions

Zoe Kean

NEWSOUTH

UNSW Press acknowledges the Bedegal people, the Traditional Owners of the unceded territory on which the Randwick and Kensington campuses of UNSW are situated, and recognises the continuing connection to Country and culture. We pay our respects to Bedegal Elders past and present.

A NewSouth book

Published by
NewSouth Publishing
University of New South Wales Press Ltd
University of New South Wales
Sydney NSW 2052
AUSTRALIA
https://unsw.press/

First published 2024

10 9 8 7 6 5 4 3 2 1

NATIONAL LIBRARY OF AUSTRALIA

A catalogue record for this book is available from the National Library of Australia

ISBN 9781742238104 (paperback)
9781761179037 (ebook)
9781761178139 (ePDF)

Internal design Josephine Pajor-Markus
Cover design Akiko Chan
Cover illustration Epaulette Shark by Marinewise.com.au
Printer Griffin Press

This book is printed on paper using fibre supplied from plantation or sustainably managed forests.

UNSW
SYDNEY

Contents

Most of these words were written
at the base of kunanyi and at the
mouth of timtumili minanya,
the land of the Muwinina people,
which was stolen, never ceded.

This book is dedicated with love
to Mia and Tommy.

Introduction

Do you ever stop and question why and how we have evolved to be the way we are? Survival is important to us, so why are we willing to risk our lives for those we care about? In a world where some species reproduce without sex, why do we need to find a partner to reproduce? Why do we fall in love, and is there a purpose in pleasure? Why are there males and females? Is life really that binary? Why do we get cancer? Why do we age, get drunk – even though it's bad for us – or spend a third of our lives asleep? And why did we evolve consciousness and develop rich inner lives?

In contemplating evolution, we see the astonishing adaptability and persistence of life. Life on Earth has survived meteorite strikes, ice ages and continent-wide volcanic events. We, and the living forms we share this planet with, are the direct descendants of the survivors of those cataclysms. How has this history shaped us? Can learning about these feats of adaptation help us to live a better life? It's these questions that have brought me to Gutharragudu/Shark Bay in Western Australia, a place where red desert sand meets the sea.

I'm starting my investigation at Gutharragudu because some of the secrets of the beginnings of life are held by creatures quietly photosynthesising just offshore.

I flew here on a small regional plane from Perth, having already crossed the continent from my home in lutruwita/Tasmania. The plane had two propellers and its engines roared as we travelled north. When we commenced our bumpy descent and burst through the bulbous grey clouds, the vast glittering bay revealed itself – a shining patchwork of luminous blue streaked with dark patches of seagrass meadows waving below the water. In the local Malgana language, Gutharragudu means 'two waters', which describes how the 23 000-square-kilometre bay is split down the middle by the red dunes of the François Peron Peninsula.

The light colour of the water is the first clue that Gutharragudu is special. The water in the bay is not deep, and this is particularly true in an ultra-shallow pocket called Hamelin Pool. The shape of the bay, combined with a sediment wall caused by the seagrass, means that water flows into the pool at a higher rate than it flows out. The beating heat of Western Australia's sunny days causes the trapped water to evaporate fast. These factors combine to make the massive pool, which is 1270 kilometres square, almost twice as salty as the ocean. Hypersalinity is bad news for most species – as anyone who has had a salty meal can attest, it's very dehydrating. In this unique environment, only the most salt-tolerant make it. The extreme conditions have allowed ancient forms of life, rare in our current world, to survive and thrive, providing a glimpse into what life on Earth may have been like billions of years ago.

Cyanobacteria exist in much of the ocean, but are greedily gobbled up by sea snails, so they never get the chance to accumulate. But snails cannot hack the salinity of Hamelin Pool, which gives these tiny single-cell organisms the opportunity to collectively achieve something incredible – to build stone structures that can grow to over a metre tall. These are called microbialites. Microbialites have different names depending on exactly how they are formed – they might be known as thrombolites or stromatolites. Both are found at

Hamelin Pool, but here they're generally called stromatolites. These rare creations are found in only a few places in the world outside of Western Australia, including a reef in the Bahamas and on the bottom of some Antarctic lakes.

The stromatolites' domain stretches for kilometres, creating uncanny reefs along Hamelin Pool's remote beaches. Access to them is strictly limited; if you want to swim among them, you need to go with a trained guide. 'I usually take astronauts and astronomers out here,' explains our guide Luke, as a small group of us travel over the red dirt track towards the beach. If life is ever found on other planets, quite likely it would look like these stromatolites.

I can certainly imagine this place on a planet in a galaxy far, far away. The sun is out as we drive from red desert to the glaringly white beach. Hopping barefooted out of the four-wheel-drive, I'm surprised by the crunch between my toes. My feet are met not by sand, but countless fingernail-sized seashells, metres deep. The cockle (*Fragum erugatum*) is another species that is adapted to the salty pool and has multiplied in its billions, creating long stretches of brilliant white coastline. As I look into the water, caramel brown shapes are visible in the teal shallows. Are they the stroms? Yes! With a rush of excitement I realise that I am about to get an insight into what life was like on the ancient Earth.

Luke shows us the narrow way by which we can wade in without accidentally touching any of these strange forms. Once I'm knee-deep, I launch in, float on the surface for a moment and then start gently kicking. Within seconds a group of stromatolites reveal themselves, the choppy water causing sand to swish around them like structures in a snow globe. They are about half a metre tall with bulbous tops, dimpled, and an odd grey-cream colour. I bob above them, oddly buoyant in the hypersaline water. The local Malgana people regard the stromatolites as their Old People (ancestors) and as I swim on, they remind me of a phalanx of stony warriors. Further out, the water clears and their shapes change, becoming smaller and flatter and forming mosaic-like patterns on the sea floor.

Each form could be thousands of years old. Geologist Erica Suosaari of the Smithsonian Natural History Museum has a long record of researching Western Australia's stromatolites. Later, she tells me over Zoom that tiny bacteria created these stone structures in two ways. One is by 'trapping and binding' sediments, like sand, that happen to wash past, creating a concrete-like substance. But they also 'precipitate' minerals from the seawater, undissolving them from the water to create limestone. Coral also uses seawater to create its stone skeletons. This means only the outer layer of the stromatolites is alive.

Some of the first life on Earth looked exactly like this. If I'd flown further north and inland, I would have arrived at an arid part of Western Australia, confusingly called North Pole, where 3.4 billion-year-old stromatolite fossils have been found. Considering the Earth is about 4.5 billion years old, these are truly ancient.

Are the organisms building the living monuments I'm swimming over anything like the tiny cells that built North Pole's precious relics? I put this to Suosaari and she explains that for many of the ancient fossils, nailing down exactly what kind of single-celled critter made them is challenging, so we can only learn about the processes these living things used to build their monuments. At Hamelin Pool, she says the stromatolites are built in a way that is 'very analogous to ancient structures ... regardless of the species, it's a process that's been happening for billions of years and it's incredible'. It is possible the earliest stromatolites were made by cells called archaea – simple cells that are subtly different to the bacteria that were foundational for the evolution of complex life.[1]

The dominant species at Hamelin Pool, a photosynthetic cyanobacteria called *Entophysalis*, is ancient. Evidence of it stretches back at least 1.8 billion years.[2] *Entophysalis* is not present in the world's other large stromatolite system in the Bahamas, and for Suosaari this makes Hamelin Pool 'the most incredible place on the planet, because you really do have this window into the ancient'.

For at least 80 per cent of the history of life on Earth, stromatolites were the most common way life presented itself – microbes were the only game in town. But then things changed. However, here in Gutharragudu they live on, emerging out of the millions of *Fragum* cockles on the sea floor. A range of small unassuming silver fish, adapted to the extreme salinity of the water, dart around them. Less common are the baby-blue jellyfish and the languid metre-long sea snakes that contain enough venom to kill dozens of people. One olive-green snake takes a break from hunting to headbutt my camera and playfully swim through my hair before ribboning off, leaving me equally awestruck and frozen with fear.

As I snorkel through this alien scene set in crystal blue, I'm struck by the unlikeliness of it all. For much of Earth's history, life existed in these relatively simple forms. But from them, and over millions of years, endless forms – most beautiful and most wonderful – have evolved.

The code to life

I don't look much like a cyanobacteria – for one, you don't need a microscope to see me. But deep in our cells we have a lot in common. Both of us are built using information stored as DNA. To get to grips with how life on earth evolves, we need to get into the nitty-gritty of Deoxyribonucleic acid, DNA to its friends. The broad strokes of evolutionary theory were understood decades before we knew about DNA.

DNA provides the instructions from which our bodies are made. Molecules called nucleic acids link together in a chain, creating mega-long complex molecules. The order of these molecules provides the information needed to create an organism – our genome.

Nucleic acids preceded life as we know it. They were most likely cooked up in a hotbed of chemical reactions in ancient Earth's numerous mineral-rich volcanic pools, though some scientists believe this happened in deep-sea vents, or even that the ingredients for life arrived on Earth via an asteroid. Either way, early Earth was a laboratory, hosting chemical reaction after chemical reaction until one day everything came together to form a cell. While this may have happened more than once, only one lineage had what it took to survive. This cell is named the last universal common ancestor or LUCA. From LUCA, everything that has ever lived evolved. From the ancient stromatolites of Western Australia to dinosaurs, deep-sea worms and humans, we are all related to LUCA.

When one cell divides into two, its DNA is replicated. Mutations can sneak in at this point if the bases – the four information points contained within the DNA – are accidentally deleted or swapped with other bases. DNA can also be scrambled by mutagens, like radioactivity. Sometimes genetic codes will undergo even more dramatic changes when new genes are added or deleted, or even more radical change occurs. Just like a beloved family recipe book, genomes are passed down the generations – if one owner cuts a recipe out, pastes in a new one from a magazine, or includes a note to add a spice, this alteration can be inherited by the next generation.

This innovation marked a massive shift in the complexity of life on this planet. DNA was crucial in the journey from LUCA to now. It is a replicable, portable, information storage system that can be described as akin to the recipe book, providing the instructions to make an organism. But it is not the chef, it does not build the organism – other processes do the actual cooking. Here's something that blows my mind: when we write a recipe book in English, we use an alphabet consisting of 26 letters. Whereas the body builds itself with variations of only four information points called bases, and computing is built from a code of just zeros and ones.

To continue with the cooking analogy, the meals the cell is cooking are proteins. The 'recipe' for each protein is a gene, which is a stretch of DNA coding for that protein. However, two versions of the same recipe book might have two slightly different recipes for the same dish. For example, a modern chocolate cake recipe might use butter instead of lard. The same can be true for two organisms. Different versions of the same gene are called alleles. Think of the gene that codes for eye colour. Just as each chocolate cake recipe still creates a chocolate cake, the eye colour gene is similar for everyone but some alleles code for blue eyes and others brown. Blue or brown, the iris is still an iris.

Evolution has many definitions and ways to conceptualise it, depending on your perspective and what aspect of evolution is being described. If you ask a geneticist, or most biologists for that matter, what evolution is, they might describe it as 'the change in frequency of a gene or allele in a population over time'. That sounds pretty abstract. But as DNA is providing the instructions for the building of different life forms, DNA is at the heart of describing how species change ... most of the time ...

So, what is pushing this change? The most famous driver of evolution is natural selection. Natural selection describes what happens when a heritable trait, usually coded by a gene, is 'selected for' because it provides an advantage. What type of advantage? The trait must help the individual who has it survive and have offspring that in turn have this trait that aides them in survival and reproduction. These traits could be anything, from size, to specific immunities, skin, scale or feather colour, personality ... take your pick. As there is always variation in a population, genes that give their owners the edge will increase, as will the trait they code for. But not all evolution is directional. A lot of change occurs due to completely random chance. This is called genetic drift, and on this very chaotic planet the effects of luck and change are not to be sneezed at. Another rather random driver of change is gene flow. This occurs when populations meet, mate and swap alleles. In the cases of gene flow and genetic drift, natural selection may, or may not, return to further influence which traits become most common.

Evolutionary change can be far more dramatic than the slow ebbing and flowing of alleles over time. As innocent as cyanobacteria seem, as I snorkel over the stromatolites at Gutharragudu, these single-celled organisms hold a dark secret. Their ancestors were responsible for the first mass extinction on planet Earth. When Earth was young, there was very little free oxygen in the atmosphere – which was all well and good as oxygen was toxic to much of the life existing at that time. But cyanobacteria release oxygen as a by-product of photosynthesis. Oxygen-producing photosynthesis made stromatolites, and oxygen floating free in the oceans changed the chemistry of the world. When the oxygen levels became too high, around 2.7 billion years ago, many tiny species went extinct, and life could have winked out at this point.

Calamity can open the door to opportunity. Much of the change that drives the diversity on our planet has occurred through slow mutations tinkering with genetic variation and then being selected for by the mechanisms of evolution. But sometimes change occurs through far more dramatic means. In this new environment, some bacteria thrived, fuelling themselves on a chemical reaction that used that abundant oxygen.

One fateful day, a cell that was not able to use oxygen – most likely an archaea cell – engulfed an oxygen-loving bacterium. The exact nature of this union is still debated, but once the oxygen-loving bacterium was inside the cell, it thrived. It made copies of itself, as bacteria do, by binary fission, meaning host cells could have many little guests. The host was able to use some of its guest's energy, turbocharging its capacities and sharing in the benefit of the oxygen. As the host cell divided, each daughter cell took some of these handy bacteria with it. Eventually the original bacteria could only survive inside another cell. This is the origin story of mitochondria, otherwise known as 'the powerhouse of the cell'. We still hold the legacy of this fruitful union in the cells of our bodies, which are fuelled by oxygen-loving mitochondria.

This story of near calamity followed by salvation makes me marvel at how unlikely, yet innovative, life on Earth has been. Small moments of chance spinning the history of our world in odd directions that finally resulted in us. Life has undergone other great changes, like the jump from single to multicellular life, the evolution of plants, fungi, animals – kingdoms of life! As well as odd, yet surprisingly useful innovations like a body plan with a distinct head and body, or behaviours like social cooperation. As each new dramatic innovation emerged, evolutionary forces like natural selection and random chance ruled its fate.

Ancient challenges in today's world

All animals that have ever lived have had to face the same challenges as we do, or similar ones. They got cancer, grew older, and for millions of years many species have consumed alcohol. They had to figure out how to find a mate, care for a loved one, and find a good place for some shut-eye. Researchers around the world are studying non-human animals to understand how they evolved in the face of these shared challenges, giving us new perspectives on age-old problems.

In this book I will introduce you to those researchers. Some slog through Tasmanian sleet, moving from wildlife trap to wildlife trap in search of Tasmanian devils that are evolving to outwit a deadly contagious cancer. I'll take you to a laboratory in northern Queensland to get acquainted with those studying the slumbering sharks whose daily rhythms are shedding light on the evolution of sleep. Researchers from around the world will share what they have learned about love and pleasure by logging the love lives of male chimpanzees and giving bunny rabbits Prozac to see if they ovulate after sex. From the best in the business we will learn about ageing elephants and the caring communities of whales. Like an elephant trundling into a mud bath, I'll wade into the murky territory of bitter, decades-long scientific debate. Finally, I'll return to Gutharragudu to consider the 'hard problem' of consciousness, asking, 'Why is it we can experience the Earth at all?'

Each question in this book meets two criteria. First, it must pose one, or several, evolutionary paradoxes. Things that, at first glance, seem like they should not have evolved. Second, the question must be deeply important to humanity and our survival. Evolutionary paradoxes do not disprove evolutionary theory, they just take more investigation to resolve – that's where the fun is.

Evolutionary science looks at our past to understand the present. But most scientists are not content to restrict their research to long ago – rather, they want to use their findings to work towards a better future. If the idea of building a better world using evolutionary theory rings your alarm bells, your instincts are good. Soon after the theory of natural selection burst onto the scene, its sinister cousins, social Darwinism and eugenics, reared their ugly heads. These pseudoscientific hypotheses took the basic tenets of natural selection and sought to apply them to human populations, with tragic consequences.

Today's evolutionary biologists strongly repudiate these ideas, or at least should do if they are worth their salt. The proponents of eugenics tend to see life on Earth as a hierarchy, and conveniently place themselves at the top. Not only is this wishful thinking on the part of a socially privileged few, it also misunderstands how natural selection and other drivers of evolution work. In modern life we are often pursuing a goal – it might be a promotion, buying a house, or even finishing the manuscript of a non-fiction masterpiece about evolution. So, when we see how life on Earth changes to meet the challenges of its time, it's tempting to think of life moving towards a goal, an end point, or a perfect specimen. But evolution is not goals driven. Species change over time, adapting to changing conditions, but they do not necessarily become more 'advanced'. There is no end game in evolution.

Explicit eugenic ideas have fuelled social policy with devastating consequences in many countries around the world. This includes mass sterilisation in countries as diverse as the United States, China and Italy. In Australia it was part of the justification for the policies that led to the Stolen Generations, where Aboriginal children were forcibly removed from their families, leaving a legacy that is still causing pain today. Cruel ideas can be hard to squash and I examine how people's biases can poison the pool of research.

Evolutionary biology can be used in pursuit of a better future without being sinister. How? Researchers I spoke to are using the principles of evolutionary biology to better treat cancer, to understand ageing and to help people get a good night's sleep. Our ancient past gives us a window into understanding why we use alcohol in excess. An evolutionary frame can inspire new responses to these age-old behaviours. Taking this view also helps us understand and celebrate ourselves better, revealing a far richer and more multilayered story about the gendered and sexual history of our forebears and the other species we share the planet with.

*

I've always found comfort in the resilience, complexity, and splendour of the natural world. As a kid I developed a deep love for nature and science. I had pet frogs and loved watching them change from tadpoles to adults, and I spent many hours using a plastic bag to hunt backyard flies to feed them. I observed caterpillars form chrysalises, learned to propagate cacti and grew mushrooms in my wardrobe. As an adult I've studied science and become a science journalist. This has given me the great privilege of being able to meet evolutionary biologists from around the world and to quiz them on their research.

It's an exciting time in evolutionary biology; change is in the air. The theory of natural selection was launched 165 years ago and, considering the scope of history, that was the mere blink of an eye ago. Since then, the field has expanded, creating its own taxonomy of disciplines. People can specialise in evolutionary psychology, evolutionary ecology, evolutionary [add whichever field of biology you like].

But it's not just the areas of study that are diversifying, it's the scientists themselves. With change comes growth and new ideas are challenging decades of dogma. I want to dive into these new ideas and provide a window into some of the most exciting thinking currently making waves in the field. The scope is global but deeply influenced by my love for the island I call home, lutruwita/Tasmania, and the many hours I spend walking in its forests and swimming in its crisp waters.

The answers I have found are intriguing, stirring, and sometimes still developing. So, come and join me on the adventure and let's see if we can find out why we *are* like this.

CHAPTER 1
Why do we care?

Dusk was falling when marine biologist Sam Thalmann arrived on the scene of the largest mass whale stranding to be recorded in Australia. On the morning of 21 September 2020, he and his colleagues at the Department of Natural Resources and Environment Tasmania got the call to say there were long-finned pilot whales in trouble at Macquarie Harbour. On a good day, this massive, tannin-stained body of water on Tasmania's rugged west coast is a four-and-a-half-hour drive from the capital, Hobart. So time was tight as Thalmann and a team of highly trained whale rescuers mobilised and hit the road. As the day wore on, the estimated number of whales kept increasing. This was going to be a huge operation, dragging out over days and requiring scores of workers from multiple agencies to put their lives on hold to save the whales.

Long-finned pilot whales are big animals. Females can grow to 6 metres and males can exceed 7 metres. That's a lot bigger than your average dolphin but smaller than a humpback. However, as Thalmann assessed the situation, the whales looked small. A total of 270 animals were stuck on a distant sandbar. A channel of deep, dark water separated the scientists from the cetaceans. Some of the animals had already died, and the survivors would have to wait till morning as the low light made conditions too dangerous to start the rescue.

Tasmania is a hotspot for whale strandings. The island state is bathed in rich, cool southern waters that are full of food for open ocean, deep water specialists like sperm whales and long-finned pilot whales. There is no agreement as to why whales strand, but certain quirks of geography can up the odds. Sperm and pilot whales navigate using echolocation – they make sounds and assess their environment by interpreting how those sounds bounce back. Tasmania has many long sandy crescent-shaped bays with gently sloping seafloors. These are very difficult environments to sonically assess. Over the decades Macquarie Harbour has caught its fair share of sperm whales and long-finned pilot whales as it has a deadly mix of sucking currents, long sandbars, and a long sandy beach to its north.

The team used the evening to plan the rescue attempt. Despite the difficulty, they knew they could mount a response. This had not always been the case. 'In the early days,' Thalmann tells me, 'it was thought that strandings were a bit of a helpless situation.' He attended his first mass stranding in 1990 – a 'small one' of about 30 pilot whales that had run ashore at larapuna/ Eddystone Point, north-east Tasmania. Thalmann remembers the deep care he felt for the five animals that were still alive when he reached the beach. 'You can hear their communication calls, trying to seek each other out … There is a lot that you connect with.' That effort ended tragically, he recalls: 'There was some deaths along the way … We had three that we cared for till about two in the morning and then it was decided that they would be euthanised … It was definitely intense.'

Since the 1990s, techniques to save stranded whales have improved dramatically. In 2019, I attended a whale rescue training session for volunteers in southern Tasmania. I was making a radio documentary and recorded the volunteers as they worked tirelessly to save a custom-made water-filled male pilot whale named Mark. As whales, like us, breathe air, they can be kept alive on the beach or in the shallows for days if they are given appropriate first aid. The volunteers manoeuvred Mark, who weighed three tonnes, using reinforced 'whale mats'. Gravity is the enemy of whales, as their body weight slowly crushes them, but by directing his head towards the sand dunes and his tail to the waves, the slope of the beach could be used to take pressure off Mark's internal organs. Sheets along his back prevented sunburn and vollies ran back and forth with buckets of water to keep the inflatable whale cool.

It took more than 20 people to manoeuvre Mark onto his whale mat and hoist him to the water. As whales get sore and stiff on the beach, he had to be exercised before release. The wetsuited team waded him through shallow water, trying not to get knocked over by waves and avoiding his tail and flippers. In the shallows the fake whale could limber up, so when he was released, he could swim to safety – being pretend, he was never granted liberty, just hauled ashore and deflated.

Even sperm whales, who are far larger than pilot whales, have been saved after stranding in large numbers in Tasmania. All this training and thought has done a lot to improve outcomes for stranded whales. However, a large factor increasing the success of these missions has been an improved appreciation of how much these whales care about each other. Thalmann explains, 'You can't just turn one around and face it to the sea and let it go. That animal will return to the beached pod straight away.'

The heavier an animal, the harder it is for its organs to work correctly when they are not supported by water. 'That is unfortunate,' says Thalmann, 'because a lot of the leadership group, the [older, larger] females that dictate the movement of the pod, die quickly.' One of the first tasks he and his team have when they get to a stranding is to identify candidates that will be able to help other whales out. Those heavier, more at-risk whales will then be stabilised, often by facing them away from the water to the natural slope of the shore to help take pressure off their organs, just as the volunteers had practised with Mark.

When they are stable, older females are often paired with each other or with juveniles for release. Genetic testing has shown that mothers and calves do not necessarily strand near to each other – but even so, this strategy seems to work. Thalmann says, 'If they release [them] together, they are already starting to form a pod.' Rescuers will often wait till they have everything set up to refloat multiple whales in quick succession to give each individual more reason to stay out.

Sadly, some whales will restrand. This can happen if they are disorientated or very unwell. But sometimes they seem to have other reasons to get to shore. Some whales 'move in a fast directed and purposeful way back to the beach. Those animals we would say have a stronger tie to maybe a missing [unit] member from that stranding site.'

The whales' strong social bonds are believed to be a big part of why they strand, alongside the sonic factors mentioned earlier. It's thought that as these events get going, the distress calls of the first few stranders bring others in. While the exact causes of strandings are still in the realms of scientific mystery, what is clear is that understanding whales' social dynamics is key to not letting them restrand

It is not just the whales' welfare that Thalmann must worry about. The distress is contagious and humans on the scene can put themselves in danger. Working with large animals is inherently hazardous. Rescues can last over a week, so emotional and physical fatigue set in. In Tasmania's cold water, hypothermia is an ever-present danger. People become so fixated on their mission that a buddy system and regular check-ins by safety officers are needed. Macquarie Harbour was too dangerous for volunteers, but even seasoned workers had to keep a close eye on each other's safety. 'People lose the capacity to self-regulate,' Thalmann says. It is not easy, and grief counselling is offered to volunteers and workers alike after an event.

This is all very strange from an evolutionary perspective. If the aim of the game is passing genes down the generations, why would whales beach themselves when they hear their fellows in distress? And why would humans – a very distantly related species – care so much about the cetaceans that they are willing to put their lives on the line for them? If nature, as Tennyson wrote, is 'red in tooth and claw', why do we care so much about each other?

On the beach, both humans and whales are being driven to extreme acts of altruism. An evolutionary biologist would say you are being altruistic if you perform an action that increases the chance of another individual successfully reproducing at the cost of your own reproductive potential.

Altruism, with its implied cost to the altruist, is a useful construct in evolutionary biology as costs and benefits can be weighed and measured scientifically, but this approach can also be limiting. It is very easy to get tangled in sticky philosophical and moral webs, worrying that if we uncover hidden benefits to the altruist, these will undermine the whole thing. For this reason, I also like to consider ideas of care, collaboration and kindness when thinking about these topics, as these ideas can be considered and valued without the idea of benefit to the carer undermining their worth. Altruism is just the action, but care is the feeling *and* the action. You can care for a whale by looking after it, but when you care about it, that is something you feel.

The 1960s, 1970s and the selfish gene revolution

The swinging sixties and the daring, individualistic seventies were revolutionary times. As long-haired pop stars strutted their stuff, introducing audacious psychedelic style to the stage, a set of equally ambitious young radicals burst onto the scientific scene. Famed evolutionary biologist Richard Dawkins may have occasionally bobbed his head along to John, Paul, George and Ringo, but the super group that really got him going was a cohort of evolutionary biologists: George Williams, William Hamilton (WD Hamilton), Robert Trivers and the slightly more senior John Maynard Smith. Unlike the Beatles, these men are not household names, but to some evolutionary biologists they may as well be rock stars. It was their ideas that Dawkins popularised in *The Selfish Gene*, which was itself a hit. The book had a massive impact on altruism studies, for good and ill. This chapter will, in part, be in conversation with it.

Just as the music of the 1960s would not have been possible without the youth culture and musical influences of the 1940s and 1950s, these scientific rebels were building on the groundwork of the previous few decades. As evolutionary biology was picking up steam in the late 19th and early 20th centuries, exactly how traits like brown or blue eyes passed down from parents was unclear. Enter genetics. At the start of the 20th century a gene was considered the 'basic unit of heredity' and the means by which specific traits passed down generations. Evolutionary biologists came to realise that we all had two sets of complementary genes, with each parent contributing half of the full set. We have met the gene before as a recipe that codes for a protein, which is a slightly different definition, but scientists like to keep us on our toes.

Impressively, this paradigm shift occurred before the structure of DNA was understood in the 1950s. The bringing together of genetics and evolutionary theory was game changing, turbocharging the quality of science. This moment is called the 'modern synthesis', and it dramatically shifted the focus from individual animals or groups of animals to the gene. The stage was set for Dawkins' rock stars.

George Williams zeroed in on the idea that natural selection acted on genes and genes alone in his 1966 book *Adaptation and Natural Selection*. Before the modern synthesis, evolution was often thought of as acting on the individual. So, if an individual has a trait – say a long tail – and that trait is heritable and helps them survive and reproduce, that trait will be passed down the generations and increase in the population. Williams coined the term the 'gene's-eye view' to advocate for a zooming-in, where it is not the trait (the long tail) that is selected but the gene (the long tail gene) that causes it. This gene is then selected for over and over again within different genetic backgrounds – in other words, in individuals. Evolution could then be understood in terms of the change in the density of gene variants, or alleles, in a population. This was a huge contribution – although later in life Williams wound the idea back a bit, suggesting it could explain a lot, but not all, of evolution.

The Selfish Gene took the gene's-eye view mainstream. As Dawkins himself admits in the 1989 edition, *The Selfish Gene* is a confusing title as the book is not about genes that encode for selfishness. Genes are just stretches of DNA, they do not have a goal. A gene is passed on if it encodes for things that make reproduction and survival more likely. Therefore, Dawkins views individual organisms as 'survival machines' evolving to act 'as if' they are trying to launch the genes within them to the next generation.

Still, in taking this view it makes sense that selfish genes would result in selfish individuals as any gene that made self-sacrificing behaviour more likely would be selected against. Dawkins writes:

> ... if you look at the way that natural selection works, it seems to follow that anything that has evolved by natural selection should be selfish ... If we find our expectation is wrong, if we observe that human behaviour is truly altruistic, then we will be faced by something puzzling, something that needs explaining.[1]

Dawkins sees another force waiting to squash altruism whenever it arises – cheaters. He puts it like this: 'Any altruistic system is inherently unstable, because it is open to abuse by selfish individuals, ready to exploit it.'

The moment a 'cheating gene' comes along, creating selfish individuals, the cheater has an advantage and can spread its genes to the next generation. Yet altruism does exist. Beyond humans and whales, it is also in wild boar, elephants, wolves, African wild dogs and more. Colony-dwelling ants, termites and bees take altruism to an extreme with whole casts of workers giving up their fertility to serve the group. So how can altruism evolve and remain in a population if we take the gene's eye view?

Let's imagine a fictitious species of bear, the care bear (*Ursus cuddlyi*). This bear is badly named, as mothers are neglectful compared to other bears, ditching their cubs at the first sign of a predator. But one day a gene mutation occurs, and a cub is born whose brain has an increased sensitivity to oxytocin, the hormone that helps build social bonds, including the mother-infant bond. This cub grows up to become an attentive mother, and when her cubs are threatened, she fights back, saving them. As a result, she is a very successful parent, raising more cubs than the average care bear. The cubs that inherit the caring mutation are also more successful. Over generations the caring gene variant spreads through the population and soon all care bears are protective mums. It's a success story for the caring gene that is now more common.

Extended forms of care are trickier to explain. William Hamilton made this point when he was a graduate student, writing:

> If natural selection followed the classical models exclusively, species would not show any behaviour more positively social than the coming together of the sexes and parental care.[2]

By thinking about genes, rather than individuals, Hamilton was able to put self-sacrifice back into the framework of natural selection – as long as self-sacrifice remained in the family.

How could this operate? Imagine a flock of related honeyeaters. Within the flock some birds have a gene that makes them give an alarm call when they see predators, such as a goshawk. The bird who gives the alarm risks drawing the attention of the predators but allows the other birds to escape. Even if an occasional 'alarm caller' honeyeater never reproduces, or has fewer offspring than others because it is goshawk dinner, the 'alarm-call genes' live on in the relatives saved – so the self-sacrificial behaviour will survive.

Hamilton quantified when this could work by constructing a simple maths equation that researchers can use or modify to see if this process is at play in the system they are studying. If the gene that costs an individual reproductive opportunities benefits its kin via increased reproductive output, the maths works. As another famous biologist, JBS Haldane, once crudely put it: 'I would lay down my life for two brothers or eight cousins.' Not a method of moral accounting I can get behind.

Hamilton formalised this view, calling it 'inclusive fitness', as an individual's evolutionary fitness – the genes they pass on – is inclusive of their relatives, although many people call it 'Hamilton's rule' or 'kin selection' – a trait selected for by kin helping kin. Kin selection is used by some to explain why, in many ant species, workers give up their reproductive potential. Nevertheless, by serving the queen, their mother, they are perpetuating their own genetic legacy.

Kin selection is one of the most influential ideas in the study of evolution and behaviour, although there are a minority who dismiss the idea, or at least our capacity to observe it.[3] Kin selection can have a limited capacity to explain altruism, especially when it occurs outside the family. Despite this, most researchers believe it is a force behind systems of care – sometimes.

John Maynard Smith took on the challenge of figuring out how altruistic genes could survive outside the family. In the 1950s a new, rational way of looking at the world was developed: game theory. Game theory is used to understand how 'rational actors' would behave in situations where there are two or more actors and the positive and negative outcome for each actor is influenced by the behaviour of the others. The most famous example of this thinking is the prisoner's dilemma, where the sentence of two prisoners, who cannot communicate with each other, is influenced by whether the other snitches or stays silent.

Game theory was quickly integrated into Cold War military plans and economic modelling. And Smith brought it into biology. His evolutionary take had a slight twist, however: it did not assume rational actors. Instead, he describes:

> ... a population of individuals adopting different strategies and playing contests against one another, and how successful they are determines how many kids they have, and they pass on to their children their properties [genes].[4]

In this way, selection can favour strategies that emerge 'as if' they were chosen by rational actors. Smith used this logic to describe circumstances that would result in stable states of cooperation and conflict. It is especially useful because it explains how these opposing forces can remain in a population at the same time. His logic was extended further by evolutionary biologists and is still used today.

Game theory is abstract. What if the evolution of altruism was as simple as 'I'll scratch your back if you scratch mine'? This is the idea behind reciprocal altruism, a model suggested by Robert Trivers that describes a theory of altruism that is not altruistic at all, but calculating. This idea, which is backed by maths and also invokes game theory, predicts that altruism can be selected for in contexts where altruistic acts can be expected to be reciprocated down the track – so what appears to be a sacrifice is nothing of the sort.[5]

Trivers argues this form of altruism can evolve in social species where individuals associate together over time. A key part of this model is the co-evolution of cheater detection mechanisms, so individuals can keep track of who is helping them and not get exploited. Reciprocal altruism can evolve because what at one moment might look like a sacrifice becomes a benefit in the end. This idea, too, has its critics.

Do selfish genes explain whale strandings?

Sperm whales and pilot whales live in tight-knit societies. Long-finned pilot whales tend to live in units of about ten individuals who can be of any age and sex. It is thought that members of a unit are closely related, possibly descended from one female.[6] Social units tend to be stable, but they are by no means exclusive. They regularly meet up with other units to form larger groups, so each whale has a large, dynamic social network. Their habit of hanging out in the open ocean means they are difficult to study, and much of what we know about their life is from research conducted on Northern Hemisphere populations, which might differ from southern groups.

More is known about sperm whale society. Like pilot whales, sperm whales also form small social units of around ten individuals, and those units often join to form temporary larger groups. They are chatterboxes, communicating with each other with specific patterns of clicks called codas. Like children learning to speak, calves learn codas from older whales. Which language humans speak is determined by their culture, not their genetics, and the same is true of sperm whales. Researchers have distinguished seven codas in the Pacific Ocean alone. Sperm whales that speak different languages often overlap geographically. However, only they tend to associate with whales that 'speak their language', forming distinct clans that can include 20 000 whales.[7]

Both species are long-toothed, big-brained predators. Sperm whales are true titans. Females can grow to 11 metres and males to 16. Luckily for us, both predatory species are squid-eating specialists. To hunt they must descend to the deep. Long-finned pilot whales have been recorded diving to depths of over 800 metres for up to 18 minutes.[8] Sperm whales are more extreme. Their dives can exceed 1000 metres in depth and extend for over an hour. They often dive multiple times in a day.[9] This poses a problem for mothers, as calves can't hack deep dives and it is presumably dangerous to leave young animals unsupervised in the open ocean. Enter the babysitters. In both species, adults will look after each other's young on the surface while their mothers dive for food.[10]

This cooperative practice of shared care is called 'alloparenting', and is seen in animals as diverse as humans, Australian fairy wrens and some spiders.[11] In pilot whales, alloparents can be male or female. In sperm whales, juvenile males may help out with the little ones, but as adult males live separately, it is females who do a lot of the work. Sperm whales take this duty seriously, and allomothers have even been observed providing milk to another mother's offspring. While it is not yet known if this occurs among pilot whales, what is clear is that raising the kids is something of a team effort.

Alloparenting is a less extreme form of altruism than the mass sacrifice that stranding may represent. Joana Augusto, a PhD student at Canada's Dalhousie University, investigated how the behaviour could evolve in a 2016 paper on long-finned pilot whale alloparenting.[12] Were selfish genes at play? Kin selection is a tempting explanation for the phenomenon, as whales spend most of their time in small units that tend to be made up of relatives, with aunties, uncles, grandparents or siblings looking after calves. But alloparenting is also common across units when they come together in larger groups. For Augusto, this rules out kin selection as a primary driver of the behaviour.

What about reciprocity? No dice there, either. The data used was collected from whale observations in 2009, 2010 and 2011. In that time, there was no evidence of mums swapping childcare favours. However, when a mother had her own calf, she was less likely to babysit, so maybe she repaid the favour when her parental duties calmed down, therefore in her paper Augusto is not ready to rule out a system of reciprocity playing out over a larger time scale.

Maybe there is no cost to looking after another whale calf? If so, finding an evolutionary reason to explain the sacrifice would not be necessary. Augusto suspects this is the case. I am not so convinced. Babysitting is trying, whether you are a human or a whale, and research on other species, such as meerkats, has shown alloparental duties can result in reduced foraging time.

What about sperm whales? Their cultural diversity extends beyond coda type. Clans have geographically distinct foraging patterns and systems of childcare. A 2009 paper led by Shane Gero of the Dominica Sperm Whale Project compared alloparenting in sperm whales living in the Caribbean to those living in the Sargasso Sea and found:

> In the Caribbean system, specific preferred escorts shared strong social bonds with and provided the bulk of the allocare to the calves, although all or most individuals in the group escorted the calves at some point. In contrast, in the Sargasso Sea, multiple nursing escorts provided alloparental care for the young, but overall, a smaller proportion of the group escorted the calves.

Gero and team were trying to determine how kin selection or reciprocal altruism could be operating in these distinct whale cultures. In the Caribbean, collections of whales tended to be made up of six to seven individuals, suggesting that they were units of relatives. But in the Sargasso Sea, whales tended to hang out in collections of 14, leading to confusion as to whether these were large units or groups of two units. This meant that the team was more confident in suggesting kin selection was at play for the Caribbean system, especially as each unit rarely had more than one suckling calf at a time, so short-term reciprocity is hard to achieve. Previous genetic research on a unit in the Caribbean showed that the calf's go-to babysitter was its mum's closest genetic relation. Interestingly, after paternity was considered, the babysitter was not the calf's closest relation in the group.[13]

The larger units or groups of the Sargasso Sea meant there were enough nursing mums for short-term reciprocity to come into play. The small community of babysitters in the extended group did tend to be other nursing mums, and they were spotted providing allo-nursing as well as just keeping a watchful eye over the youngsters. This could be evidence for reciprocal altruism, although in my view it could also be evidence of practicality. Taken together, Gero's team write that both kin selection and reciprocal altruism could be at play in these systems. But it's not a slam dunk. They write: 'It is possible that neither mechanism is sufficient on its own to explain the evolution and maintenance of the escorting systems observed in sperm whales.'

What about the more extreme behaviour of mass strandings? Strandings of long-finned pilot whales can involve hundreds of individuals. It has been suggested that these groups may all be descended from a single female ancestor – extended matrilines. A leading hypothesis on why strandings occur is that social bonds and 'instincts for group cohesion' selected for through kin selection can prime individuals to rush into shore to be with their relatives. Could strandings be an example of kin selection gone wrong?

DNA sequencing can give a peek into the family structures of beached whales. A study led by marine biologist Marc Oremus and colleagues tested the DNA of 490 deceased individuals from 12 strandings in Tasmania and New Zealand. In nine of the 12 strandings, multiple maternal lineages were evident. This challenged the idea that beached whales have a kin connection through an extended matriline that drives the behaviour. However, it still leaves open the possibility of more entangled kin connections having a role to play.

Oremus and colleagues expected beached calves would be found next to their mothers, or at least close relatives. DNA testing revealed this was not the case. Calves were scattered at random along the beach and not likely to be beside their mums. This suggests whales do not seem to be chasing their kids, grandkids or parents up onto the beach to try and save them. Many calves sequenced had 'missing mothers', who were not represented in the data. This could be the result of incomplete sampling, or some mums being refloated and surviving without their calves. However, there were enough missing mothers to make the researchers suspect some calves swam to shore without their mothers.

This is bizarre when you think about it. Even taking alloparenting into account, it would seem strange for so many calves, especially young ones, to be far from their mums. The related social units that whales spend most of their time in are small. However, hundreds of animals can be caught up in these events. If these groups are forming for feeding or breeding, maybe strandings are initiated by chaos and a breakdown of social cohesion.

Oremus and colleagues are sceptical that kin selection has driven the stranding behaviour, given the combined evidence that whales caught in these tragedies are not necessarily close family members and that families are split up as they strand. They write:

> If care-giving behaviour (epimeletic behaviour) is a force in the group cohesion of mass strandings, it may have evolved through alternate mechanisms, such as reciprocal altruism.[14]

They point to long-term social bonds between unrelated adults in related species such as bottlenose dolphins as examples of close connections without kin ties in cetaceans.

As to why we humans care so deeply for stranded whales, it could be the hijacking of processes developed through kin selection or reciprocal altruism. But as we are not genetic kin with whales, and they cannot repay the favour, I suspect there is something more going on.

Another point of view

The rational, maths-based selfish gene view was popular for a reason. It is an elegant and sometimes useful way to describe evolution. But, as I have found in reading the whale stranding and alloparenting literature, it has its limitations. Are there other ways altruism can evolve? *The Selfish Gene* is an acerbic drubbing of the idea that altruistic traits evolve 'for the good of the species', an idea now called 'naive group selection'. This was important work and expunged a lot of woolly thinking from the field. But Dawkins and his subsequent followers took this argument one step further, arguing that the gene, and only the gene, could be selected for. Selection could never operate on groups.

The radical notions in *The Selfish Gene* ushered in their own orthodoxy. But from the get-go, some researchers were wary.

What if natural selection did not only act on genes? What if cells, individuals, whale pods or human villages could possess traits that were selected for *as well as* genes? This idea is called multilevel selection, which is also a possible explanation for how multicellularity evolved. Exactly how traits of a group could be selected for, rather than a gene, was explained to me in 2019 by one of the idea's leading proponents, David Sloan Wilson of Binghamton University. He used the example of an experiment performed by a researcher who used chickens to test his idea.

The chook breeder, animal scientist William Muir, set up two experiments involving hens in cases that were essentially small agricultural cages. In one experiment he selected the most productive egg-layers from each case to breed from. This would seem a sure-fire way to artificially select generations of case-dwelling super producers. Not so.

Wilson told me: 'If you select the most productive individual in a group, you're selecting the biggest bully. And yes, that's highly heritable. And so after five generations, you breed a nation of psychopaths, and they're murdering each other and plucking each other's feathers, but they're not laying eggs.'[15]

In the second experiment, hens from the most productive cases were selected to breed the next generations, but this time these were 'the most docile and cooperative hens who did not interfere with each other'. Wilson explained to me that after generations picking productive groups rather than individuals, egg-laying actually increased. Wilsons' argument is that sometimes whole cooperative groups, like whale pods or communities of humans, can be so cohesive that the properties of the entire group are selected for. This does not preclude gene-level selection – multilevel selection allows for selection to occur at different levels of organisation at the same time. Just like the selfish gene ideas, multilevel selection can be described mathematically with a handy equation called the Price equation. This is used by evolutionary biologists to figure out at which level selection is acting most strongly. This new group selection is a departure from naive group selection, as organisms are not acting 'for the good of the group' per se, they are just being selected for as a whole group.

Be warned! This is controversial territory. For decades after *The Selfish Gene*'s publication and the emergence of the new improved idea of group selection and multilevel selection, the fight was on. Scientists had their camps, accepting group selection or not, and would hurl insults at each other like 'pseudoscientist' and 'journalist' across the barricades.[16]

Psychologist and interdisciplinary evolutionary biologist Athena Aktipis of Arizona State University has been fascinated by collaboration since she was a teenager, applying evolutionary ideas of collaboration to systems as varied as pandemic response, cancer in the body and care in humans. She is part of a new generation of scientists and less invested in the fight, describing it to me as 'frankly silly, because there isn't really a contradiction between [the two ideas]'.

Instead of hurling insults, she focuses on identifying how selection is occurring at different levels. For natural selection to take place there must be variation among traits, because when everything is the same there is nothing to select for or against. Aktipis explains that to determine whether genes, individuals or groups are being selected for, she must determine whether variation is greater between groups or between individuals. When there is more variation between groups than within, this is a hint that group selection could occur. For a group to be selected it must have some level of stability. Is this stability possible outside of carefully controlled chook-based experiments?

In the paper by Augusto and colleagues that attempted to determine how alloparenting evolved in pilot whales, group selection was dismissed. This is because alloparental care occurs between members of units when larger groups meet up, and these meetings are very ephemeral. However, this might be a bit too soon to call, as much of the social lives of these whales is still unknown. The groups could interact regularly over long periods of time, so while each meeting only lasts hours or days, there could be stability over the long term. Or the altruistic behaviour might have evolved by group selection operating on smaller units. But now that the whales have evolved caring hearts, they are willing to spread kindness beyond their small units.

Sperm whales may be better candidates for this type of evolution. As different sperm whale cultures have different systems of care, if some groups were ultra-altruistic compared to others, and this provided the ultra-altruists with an advantage, they might be selected for, and ultra-altruistic groups might increase within the population. While the possibility of sperm whale multilevel selection has been floated, most of the current speculation about the evolution of care in the species centres around kin selection and reciprocal altruism.[17]

To be altruistic, you need to have other entities around to care for. This means care and collaboration are 'emergent properties' that only manifest in the context of community, Aktipis says. A strength of group selection is that it can explain how this emergent property is selected for. As sperm whales are highly cultural and learn behaviours from each other, ultra-altruism could spread via cultural evolution, with the set of behaviours being selected for rather than genes. However, if there were genetic differences causing the variance in altruism, this could also be selected for at a group level. Thinkers like Aktipis have been developing tools that further explain how and when selection at different levels can occur. It would be exciting to see what we might learn about caring among whales if cetacean researchers employ these tools in their work.

Studying the evolution of behaviour in whales, humans or ants is always going to be challenging. Maths is hard, and animals are idiosyncratic. Therefore, a level of simplification will always be needed to apply the mathematics of evolution to the chaos of life. Aktipis says that the early models describing the evolution of altruism developed by Smith, Trivers and Hamilton were very lean, so they 'partially didn't have the capacity for representing the real world in ways that made sense'. This was because the models could not track how relationships developed over time, or how the environment impacted those relationships. Instead they tracked how selfish genes would proliferate when individuals with different strategies had random interactions with each other. Even in models where some history between individuals was included, such as in reciprocal altruism, there was no room for lasting relationships.

With colleagues, Aktipis has created her own models that she believes better reflect how relationships form over time and space. Compared to the theoreticians of the 1960s, 1970s and even 1980s, she has a lot more computer power. This has allowed her to run complicated simulations that feel truer to life. These model how cooperators and cheaters interact over time and space – cheaters being those individuals that enjoy the advantages of group life without giving back, sometimes even harming the group.

In life, if we are lumped in a group with a bunch of slackers who do not pull their weight, we generally try to leave that group. The capacity to walk away was not represented in early models explaining the evolution of cooperation. Aktipis created models with 'cooperators', who always tried to work with others, and 'defectors', who would cheat. The cooperators would 'walk away' when the group or partnership was being undermined by cheaters. Groups with low or no cheating did well and remained stable, as they could reap the benefits of cooperating. This allowed for cooperative groups to be selected for, increasing the proportion of collaborators.

The capacity of collaborators to 'walk away' meant the group could now dissipate. However, when that happened all was not lost. Through repeated interactions, cooperators would find each other, self-sorting into new cooperative groups that could then be selected for. This challenges the idea that cheaters will inevitably have an advantage over altruists. If there is more to be gained by being in a group compared to being a solo operator, then cheaters will be at a disadvantage and selected against. Seems like slackers can't coast forever. Interestingly, this lens can increase the likelihood of altruism evolving under group selection as well as kin selection or reciprocal altruism.

In an interdependent world, maybe caring isn't so weird?

When I consider the ideas canvassed in *The Selfish Gene*, they seem enmeshed with the politics of its time. (Speaking of politics, more recently both Richard Dawkins and Robert Trivers were associates of Jeffrey Epstein, and in his blog Trivers has made clear that, while acknowledging Epstein's crimes, their association extended beyond Epstein's first stint behind bars.)[18] The researchers Dawkins championed were American and British. It was the time of the Cold War, collectivism was suspect, and highly competitive neoliberal economics was on the rise. Surely this would have influenced their ideas and played a role in their rapid adoption. It certainly flavoured their methods, as the scientists borrowed mathematical tools from market economics and war strategy to underpin their hypotheses. I put this to Aktipis, and she agreed that the mindset of the time would have swayed the science. She stressed this influence does not diminish the science, saying a lot of it is 'really useful', but has limitations.

Taken together, these works leave the distinct impression that conflict is the 'ancestral state' and every expression of altruism is an unlikely occurrence forged in unique circumstances. Aktipis rejects this, framing collaboration as ancient: 'Across the broader arc of the evolution of life, needs-based helping is fundamental.' When relatively simple prokaryotic cells, bacteria and archaea teamed up to form the complicated eukaryotic cell, this was collaboration. Altruism and collaboration can also be found in the jump from unicellular life – where every cell can reproduce – to multicellular life, where the reproductive potential of most cells is sacrificed to make a larger body. This isn't to say life is all sunshine and roses; raw competition exists and can be devastating.

But mathematical models and evolutionary predictions can only tell us so much. How do humans care for one another in the real world? Aktipis co-directs a cross-disciplinary anthropological endeavour called the Human Generosity Project with Lee Cronk of Rutgers University. The project has field sites around the world that study systems of care in small-scale communities. These include American ranchers in Arizona and New Mexico, marine foragers and horticulturists on Fiji's Yasawa Island, and various foraging and pastoralist communities in African countries.[19] Across these communities a pattern has emerged: help is offered in times of need, without expectation of reciprocation.

In the case of the Maasai – pastoralist people in Kenya – who gives to whom is formalised. Maasai will enter *osotual* partnerships, where partners are required to share livestock with each other in times of need. These relationships are not entered into lightly, and partners are mutually chosen. In other cultures, such relationships can be informal, like strong friendships. Systems of needs-based transfer also exist within families, where parents give to children, and in military units where 'brothers in arms' will risk their lives to save each other.[20] Needs-based giving results in a type of insurance called 'risk pooling', where everyone's survival is more likely as there is help when it is needed.

Studies of risk pooling in small-scale societies provide a radically different view of life to perspectives based on economic 'rational actor' models where each individual is trying to maximise personal benefit. When fates are intertwined, our capacity to reproduce and raise young is dependent on the welfare of others. Aktipis calls this 'positive fitness interdependence' and explains it can be a way of explaining collaborative behaviour between kin and non-kin alike.

Evolution occurs in a context, which is often a series of changing environments. Our species evolved on dangerous savannahs where resources may have been hard to come by without working together. We then dispersed over a range of tough and varied ecosystems. Many researchers suggest that the harsh environments faced by our ancestors may have created the selection pressures needed for altruism to arise. As the research undertaken by the Human Generosity Project shows, it is during adversity that the benefits of interdependence emerge. This is backed up by research and supported by computer modelling by Paul Ibbotson of the Open University and colleagues, showing that under harsh conditions, collaborative high-risk, high-payoff large game hunting is a better strategy than solo small game hunting.[21]

When researchers have observed communities during and immediately after a disaster, pro-sociality goes through the roof. In times of calamity, when people are suddenly facing hyper-hostile environments, competition is put aside and communities pull together to get each other through.

The open ocean creates its own set of evolutionary pressures that might reward altruism, social connection and care in sperm and pilot whales. Alloparenting might have allowed these whales' ancestors to extend their dives. In both species the calves take a long time to grow to adulthood and need a lot of guidance to learn the practical and cultural skills of their worlds. It seems that to understand how pilot whales and sperm whales evolved, we must understand how they formed their caring, collaborative groups.

For sperm whales, the shared care of calves is thought to have been the evolutionary driver of their complex social world. This is because their young are the nexus around which social bonds are strengthened, changed, and renewed. Whales face a unique challenge in avoiding predators like orcas or, in days gone by, humans. In the open ocean there is nowhere to hide, especially as bubs cannot dive deep. These conditions mean the defensive action of adults is at a premium. In this context care breeds complexity.

Rather than being a carefully measured exchange, or a favour doled out to kin, what if childcare in sperm whales was a social norm? This norm could be forged by generous patterns of generalised exchange. This fascinating possibility was suggested by Canadian whale biologist Shane Gero and colleagues in 2013.[22] In an unusual departure from selfish gene theory, they suggest that searching for one mechanism to explain alloparenting risks oversimplification. The researchers speculate that whales might incorporate complex ideas like morality into their systems of care, although they emphasise that data to support sperm whale moral reasoning is not yet available. If it is true that multiple selection factors, mixed with specific cultures, selected for patterns of care in whales, I would guess the human story is at least as complicated.

Saving the whales

Back at Macquarie Harbour with the stranded long-finned pilot whales, it was all systems go from the moment Thalmann woke up on the morning of 22 September 2020. Rescuers hopped into boats to access the whales stuck on the sandbar. Survivors were stabilised and young were matched up with older whales. Soon pairs of whales were being slung to the sides of boats and transported out of the harbour. After a health check from biologists, they were released.

The rescue effort took six days as animals were returned to the ocean piecemeal. Some whales were too unwell for rescue, and others restranded. But many seemingly embraced their new lease on life. Thalmann remembers watching the whales who escaped the scene swimming south. Some individuals were equipped with satellite tracking tags. From the boat, he could hear the whales calling to one another, and in the water, sound travels far. These calls let Thalmann know 'they were able to communicate' – even whales released days after the rescue mission began were able to rejoin the group that was waiting kilometres away.

Tragically, on the third day of the effort, helicopters searching the area found another 200 whales stranded in an isolated part of the harbour. They had most likely run into trouble at the same time as the whales on the sandbar and had been hidden from rescuers due to the dark tannin-stained waters. Most died before help arrived. Of the 470 whales stranded, 114 were rescued over the six days. Of the first group of whales, well over a third were saved. For the connected whale community, this was still an extraordinary loss, but a massive improvement on the 'helpless situations' of decades past. The rescue was made possible, in part, by respecting how much the whales cared about each other.

But why do we care? By the hardline rules of selfish gene theory, it seems bizarre that we would have evolved a capacity to care about any species other than our own. Yet here were people working gruelling hours in the Tasmanian rain to save these creatures, and deeply mourning the ones who passed.

Wolves raised by humans have been known to rush into bodies of water to save the two-legged members of their pack, and domestic dogs will often put their lives on the line to save their humans. Humpback whales will rescue other whales, seals, and in one case a giant mola-mola fish, from orca attacks in what could be cases of altruistic rescue. Although they may just like messing with orcas. And there are many stories of dolphins saving humans or dogs from drowning or shark attack.

Interspecies altruism is often discussed as a glitch that causes the mechanisms animals evolved to care for their kin to spill over to the wrong recipients. This view is perhaps weighed down by the baggage of kin selection and *The Selfish Gene*. Our pro-sociality gets triggered in disasters, so maybe whale strandings mimic these conditions and 'trick' our brains into caring. This type of spillover might explain the behaviour of Thalmann, his team and the volunteers, but I am not convinced it is the whole story.

Viewing their intense urge to help as 'misdirected' could be a little confused. There is a difference between how a trait is formed and what use it is put to. Even if our collaborative tendencies evolved within the tight bonds of humanity, expanding the circle of care is putting them to a new, valuable use. Traits that evolve for one purpose get co-opted for another all the time.

Considering how humans can have great empathy for many other species, albeit selectively, I doubt this is a new phenomenon. Collaborative relationships between humans and other species have sprung up around the world. Just consider our millennia-long association with dogs and other domesticated animals. Even very young children will go out of their way to make pups happy. These relationships do not require domestication. Historically, the Thaua people of the Yuin nation on what is now the south coast of New South Wales had a respectful relationship with orcas and hunted with them, the two species teaming up to hunt larger whales like humpbacks.

Multiple forager peoples in Africa have a relationship with a bird called the honeyguide, who responds to calls made by humans, that vary by region. After being summoned, the bird will lead the caller to a beehive. The handy human can then extract the honey, always leaving some for the honeyguide, a win-win as the birds cannot access honey without help. These relationships have caused some evolutionary biologists to suggest that our capacity to care for and collaborate with other animals has given humans a survival edge through time.[23]

Our existence still depends on the other species we share the planet with. So figuring out how to tap into our urge to care, beyond one-off rescue missions, is an urgent project. I think Aktipis's view of needs-based giving and fitness interdependence is a useful frame here. She argues that our intertwined fates helped cooperation evolve, but also continue to benefit us. While not dismissing family as an important part of the story of care, she hopes that the circle of care could be expanded through focusing on fitness interdependence, writing that it 'could help us to see past our intuition that genetic relatives are a privileged class of individuals when it comes to the benefits of cooperating'. Aktipis is discussing humans here, but there are ways this view can be expanded to include other species.

Aktipis's ideas come from studying how systems of care operate in real-world communities. Many traditions around the world centre on ideas of reciprocity – to other species, to each other and to the land itself. This is not tit-for-tat-style accounting, but a needs-based system of care. Selfish gene theory arose at a time and place when individualism and maximising self-interest were having a cultural moment. The work contained some useful frames for understanding evolution, like kin selection, but perhaps reinforced a world view that cast humans in an antagonistic role against each other and the rest of nature.

In discussions of selective pressure for collaboration and altruism, it can be easy to forget what is being selected for. In many cases it is the capacity to care for the welfare of others. Social bonds are complicated and are mediated by the body's capacity to produce and respond to certain hormones and deepened behaviours that nurture this bond. While we humans can collaborate with those we do not like, care and love for a person intensifies the urge to do so.

Many researchers shy away from using terms like 'grief' or 'love' to describe animal emotions – there is a deep scientific wariness of ascribing human emotions to other species. However, outright dismissal of the idea that other species have these feelings limits the kind of questions researchers can ask. Scientists who have dared to go there have found evidence suggesting elephants, crows and chimpanzees have strong reactions to death.

Wild elephants are fascinated by elephant corpses. They will circle the body, investigate it with their trunks and, if enough time has passed, even pick up and cradle bones. As they do this, glands on the sides of their heads that leak in times of high emotion often start to pour. This will happen even when the dead elephant was from a different herd – although elephants, like sperm and pilot whales, do have relationships beyond their immediate unit, so it is possible they are reacting to the bones of friends.[24]

Zoos provide controlled environments for researchers to study animals' reactions to the deaths of individuals with whom they were closely bonded. After the death of two elephant calves at the Chester Zoo in the UK, the herd seemed to turn to each other for support, interacting with each other more. Whereas after an adult female died, they became withdrawn. Maybe, just as with human grief, loss can hit differently for elephants depending on who has died.[25]

The death or possible stillbirth of a newborn chimpanzee at the Royal Burgers' Zoo in the Netherlands gave researchers an opportunity to see how chimpanzees responded to a troop member's bereavement. The social interactions of this troop are assiduously monitored, so even subtle changes are picked up. The troop's change in behaviour towards the mourning mother, Moni, was not subtle, however. They showed a huge increase in 'affiliative behaviours', such as embracing, kissing and mouth touching in the month after her loss.[26]

Moni received extra attention from across the chimpanzee community. But one individual, Tushi, was especially caring.

Tushi and Moni did not have a particularly strong relationship before the incident. But Tushi had delivered a stillborn infant 13 years earlier and spent time holding the corpse of Moni's infant before it was removed. The researchers speculated Tushi might be especially primed to understand Moni's loss, as is the case in humans where previous loss often increases empathy.

We are a long way from cracking sperm whale codas or chatting with pilot whales, so we cannot ask them how they feel about the members of their communities. But it is thought they also mourn their dead. Both species have been seen carrying dead calves long after they have passed.[27] This behaviour is often thought to be part of the grieving process. While it cannot be validated scientifically, I have been told by more than one whale biologist that they believe these species feel love more strongly than humans.

The idea that altruism and care can evolve through natural selection can really freak people out. That acts of kindness might have been selected for because they provided a survival advantage can lead to concern that this benefit undermines the act. An article about altruistic humpback whales published in *The Conversation* claims, 'It can be altruism in the individual case, but it is ultimately driven by self-interest'.[28] I am not convinced by this, as it gives too much agency to the genes and too little to the individuals carrying them. If a gene makes it more likely for an individual to perform a selfless act, and that gene increases in the population as a result, the act itself is not undermined.

When we give, and it feels good, we are experiencing this legacy. Marine biologist Sam Thalmann explains that when people see freed whales swim into deep waters, they get an 'endorphin rush and feeling of wellbeing … it's a tangible thing'. Their bodies are being rewarded for their act of care.

Expressions of love and grief, common by-products of natural selection for care, are not cheapened by their evolutionary history. Those emotions may have been forged by selection, but they are real, intense, and meaningful. In animals as diverse as whales, wolves, and humans, our survival has depended on acts of kindness. The very fact we care for each other is because we are the descendants of a long line of carers who gave in times of need to the benefit of all, and that is a proud history.

CHAPTER 2
Why do we have sex?

I enter the insects' lair: a series of humid rooms packed to the rafters with wire shelving, which Professor Russell Bonduriansky and PhD student Daniela Wilner guide me through. Each shelf sports plastic tubes with a stick insect or two inside. The space is illuminated with a sickly yellow light, making this already strange sight truly uncanny. We are at the Bonduriansky Lab at the University of New South Wales, where evolutionary biologist Bonduriansky and his team use insects to study sex and ageing.

'Do you want to hold one?' Wilner asks.

Before I know it a bizarre and beautiful creature crawls onto my hand. She is long, skinny, and a deep green that morphs to blue on her legs. I lift her up to my face to get a better look, and Wilner warns me to be careful. When threatened, these insects spray a peppermint-scented chemical into their aggressor's eyes. Good to know.

Their ballistic abilities gave the species their innocent-sounding name: the peppermint stick insect (*Megacrania batesii*). But it was not this skill that brought me to the lab, it was their sex lives. Many female stick insects can reproduce without having sex, which results in female-only offspring. When they mate with males, the young can be male or female. The fascinating thing about the peppermint stick insect is that they have all-female populations that have completely ditched sex for generations, as well as mixed-sex populations that still copulate. The populations are often very close to each other on a map, creating what Bonduriansky describes as a 'mosaic pattern'. Despite having small wings, the species does not fly or even walk very far. In one case there is a sexual and an asexual population on either side of a road in the Daintree Rainforest.

The strange blue-green creature hypnotically waving her front legs at me was part of an asexual population now living in the lab. She comes from a long line of females who laid eggs without having to get laid themselves. After I get a photo with her, she is returned to her tube with a tasty foliage snack, and we make our way to another room, where the sexual insects live.

Looking into the stick insect 'sex tube' it takes me a while to see the male. He has clambered onto the female's back and is clinging there like an elongated backpack. Hot.

In humans, and in animals in general, we tend to think about sex as a behaviour. The coming together of two (or more) creatures in passionate congress. But as far as sexual reproduction is concerned, the behaviour is just a means to an end, the joining of sex cells, which for humans means sperm and egg. Fertilisation – when two sex cells combine their genetic material and become one cell – is one part of the sexual cycle. It is how we inherit genes from both parents. People often have multiple children with the same partner, but their kids are not genetic clones of each other unless they are identical twins. In fact, full siblings can vary wildly. This is due to the other part of the sexual cycle, meiosis. This is when prospective parents mix and match the genetic codes they inherited from *their* parents to make unique sex cells.

Sexual reproduction is an almost universal trait. Animals do it, plants do it and fungi do it – talk about peer pressure. With sex being so common, I assumed we knew exactly why and how the sexual cycle evolved. You can imagine my surprise when I discovered that this is not the case. There are lots of hypotheses as to why we have sex, and we know a lot of the story, but the specifics are still hotly debated in the scientific community. 'There's an interesting question, a big question, about why so many organisms reproduce sexually, particularly animals,' Bonduriansky explains.

Most animals have to have sex to reproduce, they are obligately sexual. Some animals are obligately asexual, having ditched sex forever. Examples of these include the New Mexico whiptail skink (*Cnemidophorus neomexicanus*). The entirely female species produces young without males. But to ovulate they still need to find a female partner and have behavioural sex. But, Bonduriansky says, 'there are very few animals where just any female can just flip back and forth' from asexual to sexual reproduction. He hopes that by studying the stick insects and their 'unstable' system, he might be able to unpick why most animals evolved to be sexual.

Sex in the birth of life

Stick insects did not invent sex, in fact, sex was on the scene long before we were animals, or even multicellular. 'Bacteria did it first,' microbiologist Ole Herman Ambur tells me when I contact him at the Oslo Metropolitan University. He explains that bacterial sex is very different from what we do. Whether it can truly be called sex is still debated by scientists, with much of the debate hinging on definitions. But let's skip that argument for now and stay with Ambur, who researches bacterial sex and evolution. He explains to me that it was in bacteria that the tools we use to slice and dice our genetics in sex were innovated.

There are certainly no candles, mood music – or, for that matter, eggs or sperm – for bacteria. Sometimes there are not even two living organisms. To understand the sex life of bacteria we must zoom into the bacteria themselves and contemplate their DNA. Bacterial DNA is a pared back affair compared to our packed genetic landscape. Bacteria organise their DNA in a single loop that contains the genes essential for function. They also have far smaller mini loops of DNA floating around in their cells, called plasmids, which can give them extra abilities, but are rarely essential.

Bacteria, like the ones that form stromatolites, often live cheek by jowl in a very crowded environment. Their world is littered with DNA as genetic material from those large loops drifts around outside their cells. This genetic clutter is left behind when a cell 'dies' or lyses, and there is some evidence that bacteria may sometimes excrete their DNA. A bacterium can draw this extracellular DNA into itself and paste it into its own genetic loop. This is called transformation and is what Amber considers bacterial sex.

Transformation is a precise process. The new sections of DNA are not randomly plugged into the loop. It is as if pages of the recipe book containing stretches of DNA or genes are cut out and new pages with alternative versions of those same recipes are glued in. This process is called homologous recombination – things that are homologous are similar in structure, so the new gene sequence is replacing the bit of DNA it corresponds to. This means the change is conservative, not radical, and in some cases there are zero differences between the DNA. This process gives alleles a type of mobility. They can hop between individuals, and over time and generations lots of different combinations of alleles are tried.

This is an incredible skill not shared by all bacteria, so

how did it evolve? For primordial bacteria living in the ancient ocean, it was important that their DNA was in tiptop shape – as it is for us today. Broken DNA represents the damage or destruction of the recipe book providing our bodies with all the information we need to keep us alive, and may result in death. Early in the history of life, bacteria evolved a gene, Rec-A, that can repair broken DNA. This capacity to fix up DNA provided a great survival advantage but also set the groundwork for transformation, or bacterial sex, to evolve.

Transformation is not the only way bacteria swap and change DNA, they also have some processes that are far more radical. These radical forms of DNA transfer can result in bacteria rapidly sharing the genes for a new trait like antibiotic resistance around a population. They are fascinating examples of evolution in action. But Ambur does not consider these horizontal gene transfers to be sex, as they are not centred around copying and pasting DNA into existing genetic stretches, and therefore are not sexy enough for this book.

True sex

As infinitely complicated as bacteria are, their cells are simple compared to our own. Ours have oodles of DNA housed within a protective nucleus, labour division among tiny organelles, and of course mitochondria powering it all. This is because we are made of a more modern type of cell: enter the eukaryotic cell, fuelling Earth's complexity since 1.7 billion years ago. Early eukaryotes were all single-cellular, as many still are today, but some lineages would eventually become animals, plants, fungi and algae. It was in these complex, nucleated, oxygen-powered cells that sex as we know it evolved. The bacteria and archaea they evolved from are called prokaryotes.

These new complicated cells no longer relied on binary fission – bacteria's method of replication. Instead, two whole new ways of dividing evolved: mitosis and meiosis. Mitosis is essentially the process cells use to make clones of themselves. As you are reading this there will be cells in your body doing it – making new skin, replacing cells in your organs and maybe even helping you grow. The DNA in each cell is duplicated and then divided between the two cells as they split.

Mitosis is similar to what bacteria and archaea do, but their simpler cells have far less DNA than the eukaryotes to wrangle in the process. If the DNA in one human cell unfurled from its tightly organised bundle, it would be nearly 2 metres – compaction and organisation is necessary within the cell so things do not get tangled.[1] This is achieved with chromosomes.

Most humans have 46 chromosomes in most of our cells. Each one is a single mega-thread of DNA carrying unique information and organised by being wound around proteins. It's rather like looping a giant strand of wool around a series of spools. When the cell is going about its everyday business, this genetic material is allowed to be loose; in this form it's called chromatin. When it is time for a cell to split, the DNA is tightly compacted and becomes chromosomes.

The chromosomes in most of the cells in our body tell an enduring tale. Except for the X and Y chromosomes, they are each in a matching pair. Each of those matching pairs is made of one chromosome from each parent. As long as you live, no matter what happens to your parents, their chromosomes are united in matching pairs in your cells.

When your parents' bodies were making eggs and sperm, their cells had two pairs of chromosomes from their parents – your grandparents. Eggs and sperm, collectively called gametes, are different. They only have half the chromosomes of the other cells in the body – just 23. These lonely chromosomes are only paired again in fertilisation. Each of these unpartnered chromosomes has a unique mix of your grandparents' DNA that is different to other sex cells your parents made.

How did your parents blend their parents' DNA to make the sex cells that led to you? When the cell is about to divide, each chromosome finds its doppelganger. At this moment *crossing over* occurs. The DNA in these matching pairs is chopped up and sections of DNA are swapped between pairs. Each new cell now has a complement of 46 unique chromosomes. It will then split again. This time there is no DNA replication, each cell only gets one of each type of chromosome and the end result is four cells, with 23 chromosomes apiece, each one unique, except for the pesky Y chromosome, which plays by its own rules. This is why siblings from the same parents can be so varied.

Did this process of meiosis evolve from bacterial sex? Ambur the microbiologist thinks not. He believes that the processes evolved separately but both 'have the same origin'. Each process picked up the tools of DNA repair, which evolved very early in life's history, and repurposed them to mix up their DNA. Despite using the same tools, and both mixing up alleles, the results of transformation and meiosis are worlds apart.

For many researchers, the rise of meiosis was the true birth of sex. Evolutionary molecular geneticist John Logsdon has spent over 20 years investigating the origins of meiosis at the University of Iowa. I ask him why it is so different from bacterial sex. In his view bacterial transformation can be compared to when his kids used to go to a friend's house and swap Pokémon cards: 'One kid might have gained a Charizard and lost [another card].' By swapping cards one at a time, his kids could improve the quality, or at least change up, the contents of their deck. 'But what they never would do is go to their friend's house and say, "Let's just shuffle our two decks together, divide them in half and then go home."' Shuffling cards is akin to meiosis – the cutting up and mixing of genetic material that originally came from separate sources. It is a far more radical act than the bacterial piecemeal process of transformation.

So how did the complicated process of meiosis, and thus sex, appear? 'It didn't happen in one fell swoop, there had to be some kind of intermediates,' Logsdon tells me. When he first started investigating this question it was assumed that sex was not common to all eukaryotes, and that meiosis evolved sometime after the last shared ancestor of all eukaryotes. This meant those 'intermediates' might be out there, and Logsdon was eager to track them down, compare them to meiotic species and use the comparison to understand more about how the process evolved. But every time he investigated one of these supposedly sexless cells he was foiled – the species would be meiotic. Even if he did not catch them in the act, they would have the genetic material necessary for the task. He jokes to me, 'My line is, they've got the goods, they must be doing the deed.'

Not finding this ellusive intermediate has been personally frustrating for Logsdon, but it is incredibly significant to the way we understand the evolution of life on Earth. In 2005 he announced his findings in a co-authored paper.[2] It announced that the last universal eukaryotic common ancestor, the distant relative we share with trees, mushrooms and algae was … sexual. This showed sex has been with us from the start. This announcement may have not made mainstream headlines, but it was huge news for those who study the evolution of sex. Unfortunately for Logsdon, the central question of how the leap was made is still unanswered.

Multicellularity and the ultimate sacrifice

What does it mean to have a complex, multicellular body and have sex?

For a free-ranging single cell living its best life, having martinis on weekends and reproducing through cloning or meiosis, becoming multicellular is a big sacrifice. Why? When you're multicellular you have a lot of responsibilities. You must coordinate with other cells – 'Who is making proteins? When? Who is taking out the trash?' – and there is a lot to keep track of. Added to that, you cannot just divide and clone yourself whenever you want – you need be controlled about replication. Boring. In a group you must compete for and share resources. You have got to deal with your neighbours' waste products from their cellular processes. Yuck. And worst of all, you may have to make the ultimate sacrifice. If you are not lucky enough to be a sex cell, when the organism dies, your line of cells dies. Surely this is an evolutionary disaster.

Despite the sacrifices, time and again, unicellular organisms have teamed up to enter regimented, collaborative multicellular life. Eukaryotes were so unlikely that to our knowledge they only emerged once. Multicellularity, however, has risen independently at least six to 25 times, depending how you define it.[3] Some of its greatest hits include land plants, animals, and fungi. But it also includes lineages that are a bit more hipster and alternative, like red and brown algae. So, while 25 times in a few billion years is not that often, it does suggest that if, under some conditions, cells can pull off this awesome feat, teaming up has huge advantages.

To understand how eukaryotic cells got together to form multicellular life, I got in touch with evolutionary biologist

Richard Michod, who runs the Michod Lab at the University of Arizona. He and his team use everything from philosophy to mathematical modelling to understand this historic transition.

For Michod, the most interesting questions about the transition to multicellularity come from considering what the unit of selection could be.

'Sex has been reinvented [because some cells give it up], but the unit of selection, that is the real thing that has been reinvented by this transition,' Michod says. While 'at first it kind of sounds mysterious', selection can operate on multiple levels – say, the gene, the cell and the colony – at the same time. This is called multilevel selection, an idea I explored earlier when asking, Why do we care? Many scientific heavyweights have fought over whether this type of evolution is possible. Controversy aside, let's just consider how a group of cells could be selected on.

In Michod's view, multicellularity was selected for when cells living together as groups gained an advantage. This might happen when there are a lot of micro predators around, so teaming up makes cells less likely to be gobbled up. Making the disadvantages of working together worth the bother of living with others. Once the cells were a cohesive team, selection then acted on them as a whole. The better the group was cooperating, the more likely it was to survive. This continued over generations until the group became so tightly knit that it became one organism. As Michod puts it, 'The cells give up their individuality, their claim to fitness [the capacity to reproduce], and the group somehow gains fitness – that is a transition in the unit of selection.'

But what about us? How did the big complicated kingdoms of multicellular life – plants, fungi, animals and seaweed-forming algae – get here? Exactly when these branches of life evolved is debated, but the ancestors of each group probably gained an advantage by working together. The fossil of a sexually reproducing red alga estimated to be over a billion years old is thought to be the oldest evidence of multicellularity.[4] It probably took a bit longer for animals to arise. It is thought that after the Great Oxygenation Event – when free oxygen nearly killed life on Earth – the Earth's oxygen seesawed. It was only when it stabilised that animal life was truly able to get going.[5]

Just as life and later eukaryotes were so unlikely that they seem only to have arisen once, the same is true for animals. Animals are multicellular and eukaryotic. Unlike photosynthetic plants, we cannot make our own food and must consume it from the environment. Fungi do this too, but we have significant genetic differences from our fungal friends.

Most animals can move about, have organs and, well, *look* like animals. But sponges did not get this aesthetic memo. They look more like elaborate rocks, sculptures or mushrooms. They do not have a digestive tract, nervous system, organs, or many of the other attributes associated with animals. Despite this, it is possible that the first animals on Earth were sponges.[6] Another strong contender for the world's first animal is the comb jelly, a creature that at first glance looks similar to a jellyfish. Whichever came first, the sponge or the comb jelly, multicellular animal life had well and truly taken off by the Ediacaran period, which started around 635 million years ago.

On my way to Gutharragudu from Melbourne I flew over fossil beds of this era. These ancient sea floors preserved in the South Australian desert give an insight into the life of our incredibly distant ancestors. By this time, animals were getting more complex, and these remote fossil fields include the first animal we know to have had a head, the Spriggina.

If you want to get ahead in your sex life, evolving a head is a great idea. This innovation allowed sense organs to be concentrated up front, meaning creatures could sense where they were going and move efficiently in that direction. If you have sex in mind, this is a great advantage, as you can move to where the mates are. Motility, the capacity to move about, may have been selected for just that reason. But it may also have helped these early animals find a meal, so there were probably a few reasons why animals got moving.

Now, 635 million years later, animal sex has got weird. Many animals, like all mammals, reptiles and birds, are internal fertilisers. An egg receives sperm while it is safely within the body, after the sperm is delivered by the penis. The fertilised egg then develops with a lot of maternal support until an infant animal is born or an egg is hatched. Things are a little more impersonal for animals like cockles and many other shellfish. Each individual can make both sperm and eggs. Sometime between winter and spring they will spawn, releasing their sex cells, gametes, into the water. Fertilisation happens in the wilds of the current and the developing larvae drift before settling on the seafloor to live out their cockle lives. Jellyfish are more unusual still. They spend part of their life attached to the sea floor, not having sex, but producing little medusas that we recognise as jellyfish. These clonal medusas bud off their asexual sea-floor production line and drift in the ocean where they then can have sex. Weird.

But why have sex at all?

Sex is dangerous. Scientists have created an exhaustive list of reasons why sex, evolutionarily speaking, seems like a bad idea. Without sex, a species can reproduce without having to find and seduce a mate – a process that is energy intensive and could end in failure. There are risks too – sex can lead to disease or, if the act is too distracting, predation.

I see a glaring example of the costs of sex while watching a raunchy sex tape of two flies mating in the wild. I am sitting in the office of Russell Bonduriansky, continuing our conversation after touring his lab. He has been showing me videos of different insects fighting over resources. In a show of supreme confidence in his algorithm, the scientist lets the videos keep running as I pepper him with questions. Then suddenly I interject, 'Look behind you!' It is not Bonduriansky in danger, but the two Australian neriid flies on the screen. They have just been going at it and immediately post coitus are distracted, not noticing the skink creeping up on them. Quick as a flash the skink leaps on the male, gobbling him up. 'And the female did not even notice,' Bonduriansky chuckles, 'but he still got to mate, even if the only way he can get to mate is to expose himself that kind of risk.' Talk about unsafe sex.

Sex is also wasteful. Each new organism made by eukaryotic sexual reproduction only has about half of each parent's DNA. If the aim of the game – from a gene's perspective, at least – is to be passed on, sex represents a risk of being cast aside. It also slows everything down. Additionally, in a line of females that can reproduce without sex, or cells that can just clone themselves, every individual in a population can make offspring. But if you add sex into the mix, it takes two individuals to reproduce, putting the brakes on population growth. This double wastefulness is known as the 'twofold cost of sex'.

The genetic diversity that evolution acts on can be created by sex. So it can be easy to fall into the trap of thinking that sex evolved *in order to* create the diversity that would eventually lead to the abundance of life that we see today. But evolution does not have a goal in mind. If an individual has got to the point that it can reproduce, it is a survivor, and its genetics mean it is well suited to its environment.

Theoretical biologist Sally Otto uses mathematics, statistics, and experiments to try and understand what drives sex and the diversity of life on Earth at the University of British Colombia. Otto explains it is 'really risky' to break apart an individual's genes and mix them with those of another to create a combination of genes that has never been tried before.

So, what is giving sexual species the edge? A major driver of evolution is selection acting on particular alleles. An allele will increase in the population if it gives its host a better shot of reproducing; this is how it is selected. It can also get lucky and increase by random chance. But without sex, there is a problem.

These alleles are trapped! Every time an individual clones itself, there is a 'selection event' where all the genes from that individual are passed on. As Otto puts it in a 2020 review, 'In the absence of sex, alleles must rise and fall together.'[7] She tells me, 'They're stuck inside whole genomic blocks, and that's going to grind evolution to a halt soon after a period of selection.'

Sex liberates genes and alleles. How? Imagine a new mutation arising that could advantage an individual, but being stuck in a genome that has harmful mutations, or mutations that neutralise its effect. These harmful mutations might mean the potential positive effects are not great enough to increase survival and reproduction, essentially hiding the gene from evolution. As Otto says, 'Selection can't act on *a gene* in isolation of all the other genes in the genome if there's not recombination.'

Whether a species has sex is determined by its genetics. To unpick how sex might be selected for in the first place, even with all the costs associated with it, Otto has turned to mathematical models that simulate how genes can be selected for under different conditions. She's found if a gene arises that increases sex, it does not necessarily have to benefit the first generation. As long as the first generation is not so disadvantaged that it cannot survive, the benefits of sex will often show up down the track. This will happen if sex results in these later individuals having 'better combinations' of alleles. The capacity to unleash these new combos will mean the gene that allows for more sex is maintained.

What about an organism that already has great genes and winning combinations of genes that help it thrive in its environment? Could it give up sex? This would be risky. We live in a dangerous world. There are parasites and predators to avoid, environmental changes to combat and competitors for resources. An individual might be adapted well enough to their environment to reproduce, but as everything around them is also evolving, there is no guarantee the genetics that work for one generation will not be obsolete for the next generation. This is the logic behind the Red Queen Hypothesis, named after a scene in Lewis Carroll's *Through the Looking Glass* where Alice is chasing the queen and notices that both of them are running, but not moving forward. The Red Queen tells her, 'It takes all the running you can do, to keep in the same place.'

An organism is surrounded by neighbours trying to predate, parasitise or outcompete it. The neighbours will be evolving, so to outfox them the organism needs to evolve as well. In this way species put selective pressure on each other within an ecology to evolve rapidly.

When we are doing our best, and doing well, being outshone by a slightly better competitor is not easy to take. For a human, being runner-up might be a blow to the ego, but for a positive allele in a clonal population, it could be devastating. If two great traits show up in different lineages and the clones with the magnificent allele outcompete the owners of the not-too-shabby allele, the not-too-shabby allele may disappear. While it might not be as great as the magnificent allele, this is a loss for the whole population. In this way, clones are interfering with each other, so the phenomenon is called 'clonal interference'. But if sex enters the picture, and genomes are mixed and matched, some individuals might arise that have both alleles. The not-too-shabby allele is maintained, and this new combination might be even more advantageous to the individual who inherits it than having just the magnificent allele in isolation.

Sex is not all about change. Microbiologist Ambur's view of the evolution of sex is conservative. Small mutations have contributed to the fabulous array of species on Earth. He marvels at 'giraffes and African elephants, and trees and the diversity of life'. But Ambur thinks this variation can distract us from what might be a stronger force at play: conservation. This has led him to develop a 'steady at the wheel' hypothesis, where sex is maintained because it helps conserve essential genetic information. Meiosis and transformation (as in bacterial sex) do not represent wholesale changes of a genome. Instead, these processes 'combine beneficial alleles in the same genetic background'. This means the offspring, while different from their parents, are similar, and the information on how to make that species survives.

Mutations might fuel the diversity of life but for the most part they are bad news, Ambur explains. 'Three point five billion years of evolution have shaped the genes and the proteins that we are made of. So, natural selection during those 3.5 billion years made sure that they work in a pretty good way.' The chances a mutation will be helpful are low, so life needs a way to purge the mutations that do not work. In an asexual population, dodgy mutations will slowly accumulate until extinction – an idea called Muller's Ratchet. When alleles can be shuffled during sex, these mutations can be removed from a perfectly good genetic context, and replaced with a less damaging allele, maintaining individuals without the dodgy gene variant – and conserving the working gene.

This process links us to our ancient past, as many of the alleles our bodies rely on evolved millions if not billions of years ago and have been conserved. Ambur's go-to example of this is Rec-A, the gene that evolved on ancient Earth to repair broken DNA and is now used in bacterial transformation. This gene is incredibly similar to the genes our bodies use during meiosis. It has been conserved, changing little over billions of years of being cut and pasted into new genetic contexts through sex, until it got to us and to many modern bacteria. Ambur marvels, 'It is mind-boggling how conservative nature is in this regard.'

Is sex a radical act, driving change? Or is it conservative, preserving our genome against entropy across the millennia? For Otto, this is not an either-or situation. There are selection pressures being experienced constantly for conservation and for change at different degrees and at the same time. For her, reasons for sex that promote change, like the Red Queen and clonal interference, and conservative forces, like steady-at-the-wheel, can be unified under the broader idea of selective interference.

This chapter has traced the long road of sex, from ancient cells repairing DNA in a stromatolite world to tiny single cells shuffling their genes and the rise of multicellularity. It took a long time to get here, but there are species of plants and animals that have given up on sex entirely and evolved asexuality from a sexual ancestor. These include some species of tree like the King's lomatia (*Lomatia tasmanica*) in remote south-western Tasmania, which only reproduces when limbs break off, land in the earth and grow roots, or from the spread of underground roots.

What about the asexuals, the species who cannot have sex? Otto considers that for the most part they seem be 'evolutionary dead ends'. That is, if asexuality evolves, it might persist for a few thousand years or so but tends not to survive long term. So, while King's lomatia may have been cloning itself for 40 000 years, it is unlikely to lead to a large group of new species. There are some animals, like the hardy microscopic rotifers, that seem to be loving the sex-free lifestyle. But these species have other ways to mix their DNA, so they could be getting the same outcome via a different process.

Otto still has questions, though. The real head-scratcher now is why some species, like us, are locked into sex as the only form of reproduction. For many single-celled eukaryotes, sex is a rare occurrence. They will clone themselves happily until something, usually stress, signals them to have sex, when they will undergo meiosis. Many plants can happily reproduce asexually, creating clones from cuttings or runners. But these clones can still flower and reproduce sexually.

As Bonduriansky says, the capacity to flip between sexual and asexual is rare in animals. Beyond the stick insects it has been seen in only a few species. The female Komodo dragon can switch between having sex with males and reproducing and having young without males. Sharks and rays kept in captivity have achieved virgin births and in 2021 the world was wowed when genetic analysis of the hatchlings from two eggs laid by critically endangered California condors showed that no male had fathered the chicks.[8]

Being able to reproduce both sexually and asexually is a great system. When the costs of sex get too high, or there are no males around, females can still reproduce. So why lock in sex? Otto explains it could be because it is costly to maintain both systems: 'You get less efficient at the one that you don't do often, [making a] one-way street where asexual offspring are rarely made, so you get worse and worse at making them.' A one-way street might be created when there is a 'dance between mothers' and fathers' genes, creating this really intricate developmental path to make an offspring', she says. It is also possible that large complex brains are so finely tuned and vulnerable to damage that maintaining even slightly deleterious mutations for too long would be a disaster, so sex at every generation provides insurance against that.[9]

The work Bonduriansky and his students are doing seems to back the idea that animals get 'less efficient' at the type of reproduction they do less of. Wilner, the student who introduced me to the stick insect, has found 'good evidence that females from at least the southern [asexual] populations have evolved resistance to mating', Bonduriansky tells me. I ask him whether this is behavioural or physiological resistance. 'She's looking into that,' he replies. 'We don't know yet, it could be behavioural, it could be physiological or morphological, it might be that the sperm storage organs are smaller or non-functional.' The team is finding a 'maternal effect' where even one generation of asexual reproduction results in fewer of the females' eggs being fertilised when they mate.

On top of that, mating with males is costly to stick insect females. Males will fight for access to a female, and the female is exposed to the risk of injury when the victor climbs on her back – where he stays like a backpack, mating with her, weighing her down and even occasionally nibbling on her. Given this, the females are better off not having sex at all. This led Bonduriansky and his team to wonder if the all-female populations are maintained because females actively resisted mating. He speculates it is possible, in some species, that it is the very presence of males in a population – and that they are able to mate with females – that ensures their ongoing existence. This is because in species like the stick insects, asexual reproduction produces no males, but sexual reproduction creates mixed sex hatchlings.

Without the genetic mixing provided by sex, are male-free populations destined to die off? Many species that have made the full transition to a life without males seem to go extinct. For peppermint stick insects it is yet to be seen, but, Bonduriansky says, 'There is something very interesting going on allowing these asexuals to do very well.' Genetic testing has revealed that one male-free population has been thriving for hundreds, if not thousands, of years. Asexual stick insects may not mix their genes via fertilisation, but they can still keep things a little diverse. They go through meiosis, mixing and matching their own genetics, then re-join those new genetic codes inside themselves via a process called automixis. This doesn't provide as much diversity and allele liberation as the full sexual cycle, however.

Soleille Miller, a PhD student at the Bonduriansky lab, compares the health of asexual and sexual populations in different environments. The species can live in both beachy and swampy ecosystems, and her work has found that by the beach asexual populations seem to be as healthy as sexual ones, but in the swamp asexual populations were more prone to fungal infections compared to sexual ones. She says, 'It seemed like some genotypes were really well-suited for their environment and it doesn't really matter whether they were asexual or not, and some populations could be more at risk of dying out or of having lower fitness based on the environment.'

We are locked into sex. These hypotheses provide exciting hints as to why, but the answer is unknown. Meaning sex, for now, retains an element of mystery.

CHAPTER 3
Why do we have males, females and other sexes?

Fencing, combat with swords, is known as the 'elegant sport'. Combatants dodge, weave, and lunge at each other. A touch is scored when 'the tip of the blade strikes the opponent's target area with enough force to depress the point'. Of course, today's Olympic fencers play it very safe with masks and protective clothing. Under the sea the fencing is far more high stakes. Swords are replaced with penises, and competitors duck and swerve trying to avoid impregnation, all while aggressively wielding their phallus in the hope of impregnating their opponent.

The dick-slinging slugs in question? The marine flatworms, more formally known as *Pseudocerotidae*. Despite the dull-sounding name, flatworms are entrancing creatures. Flat? Yes. Wormy? Only in so much as they don't have any appendages and have a simple body plan. These beauties are often technicolour and float around their marine world looking for all the world like gaudy flying carpets. Flatworms are hermaphrodites, a word that is a slur when used to describe people, but is still used in biology to describe animals that carry both sperm and eggs. The story goes that it's 'expensive' to brew and care for fertilised eggs, so if you meet a potential mate on the reef, try and be the dad by injecting your 'cheap' sperm into them. To make matters worse, the worms do not have a vagina, so are inseminated when the penis punctures the surface of the other flatworm. A touch anywhere will score.

This is a dramatic and gory story. It is often trotted out as a perfect example of a fundamental guiding idea in biology: that sperm is cheap, and eggs are expensive, and this difference results in predictable and reliable differences between males and females. Motherhood is defined as a 'loss' and fatherhood a 'win'. But just as I imagine the surface of a marine flatworm to be, ideas around the sexes and gender are slippery and harder to pin down than one might imagine.

Sex is a biological term, whereas gender refers to the social and cultural ideas we associate with sex. Both things are important to our perception of self, and how we conceptualise them can be deeply influenced by our culture, religion, politics and general world view. This means every researcher who tackles the issue, whether they are aware of it or not, is bringing their biases to the question. The history of evolutionary biology from the Victorian era to now has been dominated by men. Mainly white, mainly from affluent backgrounds. As zoologist Lucy Cooke puts it in her book *Bitch: A revolutionary guide to sex, evolution and the female animal*, 'a sexist mythology has been baked into biology'.

These myths started millennia before the mechanics of evolution were figured out. In Ancient Greece Aristotle wrote of 'effective and active' males and 'passive' females, although exactly what he meant by that is still debated. This fallacy would have been picked up by the Victorian evolutionary thinkers who received a classical education. A view of nature with vigorous males and virtuous females nicely reinforced the role of women in their society 150 years ago as pure, soft, and lacking in true agency. Like a hapless flatworm in a marine biologist's net, these sexist mythologies have been dragged through the science from the ancient past to now.

One of the problems is that gendered myths are so seductive. After being discussed in university lecture theatres around the world – and even making an appearance on a David Attenborough documentary – the battle of the sexes implied by penis fencing in flatworms has come under question. The challenge was raised by Samantha Tong and Rene Ong of the Tropical Marine Science Institute at the National University of Singapore in a 2020 paper that extensively studied the sex lives of multiple species of Pseudocerotidae.

As they have no vaginal opening, insemination requires the skin to be stabbed. From a human perspective a violent act, but for a worm, maybe not so much. The researchers write:

> Penis fencing could also be easily mistaken as violent and physically damaging (injury on epidermis), because polyclads use their armed penis to pierce the epidermis of the mating partner in order to successfully inject sperm.[1]

After observing the worms in action, they found that 'our results showed that the act of penis fencing could be just a mating ritual, and not necessary for successful insemination'. They also note that penis fencing often results in insemination for both parties. This does not negate the idea that penis fencing could sometimes be a high-stakes game where 'the loser becomes pregnant' but suggests to me our enthusiasm to spread that narrative could have something to do with our 'sexist mythologies'. It is possible that rather than a battle, penis fencing is more like a courtly dance between suitors.

To me, the flatworms bring up three interesting questions: Why are there males and females? Yes, we know mixing genetic material can grease the wheels of evolution by liberating alleles from one generation to the next. But if everyone could get pregnant, or lay eggs, why bother separating the sexes? While hermaphroditic flatworms look identical to each other, often within species males and females look incredibly different. Why? And, if we can be so misled by flatworm sex, what else have we got wrong about males, females, and other sexes?

The rise of females and males

The producers of large gametes we call females, and the producers of small gametes we call males. This definition is simple, but like everything in biology, the rule is far less concrete than it appears to be.

Plants, animals, and red algae come in two sexes. But an incredible amount of life on Earth is well outside of the female-male system. The single-cellular algae *Chlamydomonas* and *Gonium* have gametes that are the same size, or *isogamous*. But the multicellular *Volvox* has large and small gametes, making them *anisogamous*. Isogamy is the norm for single-celled sexually reproductive species, but it is also found in some multicellular algae. Fungi cannot be boxed in by the binary. They have 'mating types' – two fungi of the same mating type cannot mate, just as two eggs or two sperm do not fuse in fertilisation. So, some fungi increase the odds of meeting an eligible partner by upping the number of these mating types, which in some ways are analogous to sexes. The count of these 'types' can spool out into the thousands.[2]

Back to males and females. Exactly how they evolved is a bit of a 'thorny question', according to evolutionary ecologist Hanna Kokko of the University of Zurich. She has spent her career investigating some of life's largest evolutionary mysteries, including the evolution of separate sexes. Anisogamy – having large and small gametes, and thus males and females – evolved long ago in some of the first multicellular creatures in the ancient ocean, meaning it is not represented in the fossil record. What do researchers do when they cannot find physical evidence of something? Turn to mathematical models, of course. Computer simulations, game theory and a raft of equations have all been thrown at this mystery.

To create useful simulations, researchers like Kokko make assumptions about what conditions were like for the species that first evolved anisogamy. Imagine an early multicellular creature that is not very mobile and might even be attached to the sea floor or a reef. When the time is right, its gametes are released into the ocean. Crucially, it creates gametes of a similar size, although there will be some random variation. Each individual has limited resources from which to make their gametes – no one is an infinite gamete machine. Fertilisation and development is tricky. Two gametes, each with half the genetic material needed to create a new life, have to meet in the water to form a zygote. Critically, for a new individual to then develop, the zygote needs enough resources.

To maximise the chance of fertilisation, the system needed to be optimised, and in many cases isogamy is anything but optimal. Kokko likens these ancestral isogamous sex cells as 'jacks of all trades but masters of none'. Small gametes of the same size run the risk of not having enough resources to provision the zygote after fertilisation. Whereas a species that only produces a few large gametes runs the risk of releasing their possible progeny into the chaos of the ocean, where they may never meet another gamete and will be fated to die alone. Never will they merge with another sex cell to form new life.

In this case, Kokko says, anisogamy is a specialisation driving 'small, abundant searchers and the less abundant, bigger, investors into the future wellbeing of this creature [that will develop]'. Producing more active-searching gametes increases the chance that a partner will be found. This takes the pressure off other gametes, allowing more to be invested in each gamete so when it is found, it is more likely to survive. As animals, plants, and red algae likely all evolved anisogamy after they diverged from each other, the trait evolved more than once.

Sex roles: is everything we thought we knew wrong?

Once anisogamy, and so males and females, evolved in plants and animals, it stuck. Packing large gametes, small gametes or both creates unique challenges for the organism that possesses them. This means within a species different selection pressures can be applied, depending on sex. This small difference is thought to be the basis of the differences between sexes. This can even be seen in hermaphroditic species. For example, flowers produce pollen, tiny grains that contain two male gametes, and ovules, which are housed in separate elaborate structures within the flower. Pollen is produced on the stamens that stick out so it can be carried away to pollinate another flower, whereas the ovule is protected in the ovary at the flower's centre.

If a trait makes it more likely that an individual reproduces or increases its number of offspring, even if it does not necessarily aid in survival or sometimes threatens it, the trait will likely increase in the population. This is called sexual selection, though exactly how sexual selection should be defined is an active and contentious debate among evolutionary biologists. That's a rabbit hole for another day.

Sexual selection can lead to sexual dimorphism when this particular trait is of greater benefit to one sex. This can influence how an animal looks – think of the flashy feathers of a peacock compared to the relatively subdued peahen – but it can also impact behaviour, from fighting over mates to who is putting in the time caring for little ones.

This idea was first floated over 150 years ago in the early days of evolutionary biology and gained hardcore empirical support in the mid-20th century. The paradigm to help researchers understand how these forces operate was locked into the scientific literature when a series of experiments on fruit flies by Angus John Bateman published in 1948[3] was dragged into the limelight by evolutionary biologist Robert Trivers (yes, the same Robert Trivers who posited the idea of 'reciprocal altruism' that we met earlier) in his influential 1972 paper 'Parental investment and sexual selection'.[4]

Fruit flies are frustrating in the kitchen but a favourite of biologists. They are relatively easy to keep in the lab and can carry mutations that lead to outlandish and heritable physical traits. By tracking these traits through the generations, a biologist can nut out the paternity and maternity of a particular fly.

This is exactly what Bateman did, creating a swingers' event where virgin females could meet and mate with males, each sporting a bizarre mutation that made their offspring easy to track. His results were compelling. Females had consistent success, with only 4 per cent not producing surviving offspring, whereas 21 per cent of the males failed to reproduce. Other males were wildly successful, fathering nearly three times as many surviving offspring as the average female.

The implication, according to Bateman and Trivers, was that as a general rule females did not stand to benefit from multiple matings, whereas males did. As females only got one shot, or at least limited shots, to pick the right fly-guy, it would pay for them to be choosy, whereas intense competition would benefit males, as it would increase the number of offspring they could have. If traits led a male to have high reproductive success, those traits would be heavily selected for. This could lead to some of the outlandish traits often seen in males, such as a deer's antlers or fancy feathers on a bird. Both scientists linked this notion back to anisogamy, suggesting that different strategies could be explained by differential investment between male sperm, which is often described as 'cheap', compared to the female's 'expensive' eggs.

These principles are so elegant they can be illustrated in the simple graph below. It has two lines, one for low investors (often males) and one for high investors (often females). The graphs y-axis represents the number of offspring ('fitness'), i.e. reproductive success, and the x-axis represents number of matings. The low investment line steadily rises as reproductive success increases with each mating. Whereas the female line eventually plateaus as her expensive eggs run out. Because of anisogamy, and compounding factors like internal fertilisation and nursing in mammals, it will usually be the females who invest more.

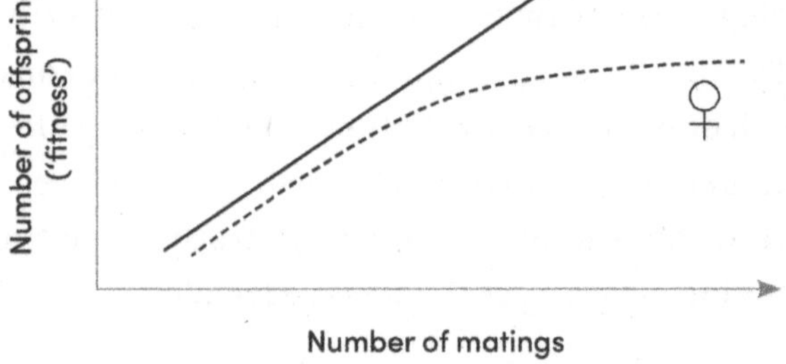

However, Trivers is careful to point out that this trend is reversed in many species that he calls 'sex role reversed'. These include animals such as seahorses, where the fathers care for the eggs in a pouch, or birds like emus or cassowaries, where dads do all the care.

The principle provides the groundwork for a few dynamics to play out:

- If the partner who invests less can have more offspring by finding other mates, there will be conflict between the sexes as to who provides the most care. For instance, a male bird might try to seduce females, instead of collecting food for the chicks.
- The sex who invests less will be in competition with other members of their sex for mates. This could explain why male kangaroos are larger than females, as large roos win more fights, allowing them to mate with more females.
- The partner who invests more should be choosier. If they only have one shot, or limited shots, to pick a mate, they had better choose well. Traits that make the opposite sex more likely to get chosen are maintained and sometimes exaggerated – imagine a peahen carefully selecting the peacock with the best tail.
- Any trait that increases the chance of 'winning' – as in having more offspring than your competitors – will increase through the population. So peacocks with ever more elaborate tails evolve.

For scientists trying to understand why males and females evolved in particular ways across species, across time and across the world, these ideas remain foundational. The results were in: males had a biological imperative to be competitive and promiscuous and females had evolved to save themselves for Mr Right. However, there is a problem here. Bateman was wrong … ish.

Biologist Zuleyma Tang-Martinez is a critical thinker. Now an emeritus professor at the University of Missouri-St Louis, she was a graduate student at the University of California, Berkeley in the late 1960s and early 1970s. This was just as the Bateman–Trivers paradigm was making waves and, she tells me, she 'bought it all hook, line and sinker'. Once she gained her doctorate and started teaching, she taught the theory without question. These days, Tang-Martinez is one of its greatest critics.

Research published in 1975 into sexually liberated red-winged blackbirds gave Tang-Martinez the first inklings that the paradigm might be more complicated than originally thought. The bird research world was scandalised when male members of this presumed monogamous species were given vasectomies but the eggs in the nests of these snipped males still hatched.[5] Clearly these females had been fertilised by someone else. Trivers had predicted that in 'monogamous' bird species, males may seek to have sex outside of the nest to spread their genetics. But the theory did not leave room for females to do the same. As such, many researchers thought that females might be getting fertilised through forced copulation, an idea that never seemed credible to Tang-Martinez.

The red-winged blackbird research primed Tang-Martinez to be suspicious of the Bateman–Trivers paradigm, but her relationship with the idea really changed in 1988. She tells me how an ornithologist named Susan Smith and her students, 'followed a population of black-capped chickadees 24/7. They were able to observe the females shortly before dawn going into the territories of other pairs and mating with males and then coming back to their own territory and nest.' This meant there was likely a benefit for females in seeking sex outside of the nest.[6] For Tang-Martinez, 'It was like a lightbulb going off in my head ... That's what started the whole cascade of my questioning, questioning, questioning.'

Since the news broke that female black-capped chickadees seek early morning quickies with their neighbours, genetic technology has advanced, showing that extra-pair lovin' is common among supposedly monogamous birds. It has also been shown that in a huge number of species, including but not limited to the redback spider, the brown trout, the black ratsnake and the yellow-pine chipmunk, a female's reproductive success is increased by multiple matings.[7]

Can females benefit from having many Mr Rights? A 17-year study of extra-pair paternity in dark-eyed juncos, a type of songbird, showed that increased mating improved reproductive success in males and females, although this research shared the same limitation as Bateman's work as it measured offspring rather than actual matings.[8] A study by Yukio Yasui of Kagawa University on female crickets showed that multiple matings made the offspring of females more resilient to extreme environmental conditions such as high temperatures and salinity, suggesting to him that mating multiply could be protective for the species as the climate changes.[9]

As exceptions to the paradigm kept popping up, Tang-Martinez decided to scrutinise Bateman's original fruit fly study. In two influential papers she argued that Bateman had not observed matings so could not assume the number of matings – a critical flaw when your research concludes that females do not benefit from multiple matings.[10]

Incredibly, for 48 years nobody replicated the Bateman study to see if it stood up to scrutiny. A repeat project was undertaken in 2012 by evolutionary biologist Patricia Gowaty and colleagues and resulted in a paper titled 'No evidence of sexual selection in a repetition of Bateman's classic study of *Drosophila melanogaster*'.[11] Remember those wacky mutations each fly had so its offspring could be counted? Well, it turns out that some of those mutations combined are lethal. So, as only surviving offspring were counted, the numbers were off. Additionally, it was found Bateman had made errors in the tedious task of counting the flies themselves. Some offspring were counted as having fathers and not mothers, a biological impossibility that inflated paternity estimates for some flies.

I was shocked to learn it had taken so long for anyone to realise that Bateman's research was so off. So were Gowaty and colleagues, writing, 'We are left wondering why earlier readers failed to spot the inferential problems with Bateman's original study.'

What now for sex roles?

It seems we are in a pickle. What do we do with research based on premises that are so shaky? Despite the sceptics, the Bateman–Trivers paradigm is based on neat logic and does explain how males and females evolved.

Distressed by the growing challenges to Bateman, a group of researchers set about trying to prove him right. Evolutionary biologist Tim Janicke of Centre d'Ecologie Fonctionnelle et Evolutive in Montpellier and colleagues analysed findings from 72 studies of 66 animal species as diverse as snails, frogs and birds. In 2016 they announced that anisogamy results in the 'conventional sex roles' predicted by Bateman's fruit flies and parental investment theory.[12] While they acknowledge that the original idea that females only benefit from one mating is outdated, they claim their analysis shows that males have more to gain by multiple mates, therefore have more variance in their reproductive success and therefore sexual selection is stronger for them, leading to the differences we see between males and females. They write: 'Sexual selection research over the last 150 years has not been carried out under false premises but instead is valid and provides a powerful explanation for differences between males and females.'

In their work, they clearly show that the dynamics sketched out by traditional sexual selection theory hold up in many contexts. But I am wary of this declarative and universalist attitude; it is hard to see what you are not expecting to see. Many of the studies incorporated in their meta-analysis were undertaken with an uncritical expectation that the dynamics expected under the Bateman–Trivers paradigm would be the key drivers in the systems studied, so these results are unsurprising. I begin scratching my head towards the end of the paper as the authors start to equivocate, noting that there are 'many exceptions' to traditional sex roles and that 'these exceptions to the rule highlight the importance of incorporating environmental conditions when interpreting animal mating systems'. Well, yes.

Even the maths backs Bateman – sometimes, under some conditions. In 2022 mathematical models showed that anisogamy would often lead to Bateman curves and sex differences under both external and internal fertilisation.[13] Crucially, the result is 'not inevitable' and changing the conditions a species finds itself in can push the curve to equality or reverse the effect. This modelling reveals a richer, more complicated landscape where sex roles have evolved.

A small minority of researchers argue that the dynamics explained by the Bateman–Trivers paradigm are never at play.

But many modern critics, such as Tang-Martinez, do not. Instead, these critics are begging that the 'paradigmatic power' of this idea be challenged in education and research. As Tang-Martinez writes, the 'ability to interpret reality with regards to male and female sexual behaviour has been compromised' because the Bateman–Trivers paradigm is often the only game in town.[14]

How are we meant to understand differences between the sexes if the way we have researched them for 150 years is flawed? Tang-Martinez recommends that when researchers investigate sex roles in a species they 'start from scratch'. She points out that in many studies of sexual selection the males get more researcher attention than the females, so she recommends researchers 'observe what the males are doing, but also observe what the females are doing. And you don't just observe for a limited period in a particular location.'

In some cases, this research will fall neatly into the paradigm, but sometimes the differences between sexes might be the result of previously ignored factors. Human males tend to be taller, more muscular, have a deeper voice, more testosterone and tend to have more body hair and more 'masculine faces'. The more pronounced these traits are, the more masculine an individual is – scientifically. It has been hypothesised that 'masculinity' has been sexually selected due to either female preference or the trait's capacity to help males win fights for females. But a recent analysis brings together data from 96 studies and finds that the only 'masculine' feature with enough data to suggest sexual selection is at play is muscularity, leaving the other traits open for an alternative hypothesis.[15]

On average, women store more fat on their bodies then men, including on their breasts and bums. Humans are unique among primates in the dramatic difference in body fat between males and females and the existence of permanently enlarged breasts. As they are generally considered alluring, a dominant argument is that breasts evolved in an example of 'reverse sexual selection'. The logic goes that females with larger breasts were more likely to be chosen by males and therefore have more offspring, but recently this idea has been challenged by multiple researchers.

The brains of human foetuses are big and hungry. It is possible that early in the human lineage when a mother had the capacity to store fat it meant her babies were more likely to be brought to term and develop well. This increase of fat could have changed the levels of sex hormones such as oestradiol, and as the glutes and the breasts are more sensitive to these hormones, this is where fat accumulation was centred. While the researchers do not rule out sexual selection having a role in increasing these traits, thinking outside this lens has allowed researchers to consider other ways sex-linked traits may have evolved.[16]

Primordial sexual conflict

For the last 150 years the study of sex roles has been the study of conflict. This conflict has been imagined as starting from the moment the sexes evolved to now. The first major modern hypothesis explaining the origin of anisogamy implies a 'battle of the sexes' from the jump. There had been earlier hypotheses in the 20th century, but they were based on ideas of how evolution works that are no longer accepted. Published in 1972, the foundational idea was developed by Geoff Parker, Robin Baker and Vic Smith and named the PBS model after their initials.[17]

The PBS model views the origin of males and females as a high-stakes, open-ocean drama that emphasises competition and cheating. Who are the warriors in this battle? Picture our ancestral isogamous species and assume there is a small variation in the size of gametes it produces. 'Proto-males' produce slightly smaller gametes than 'proto-females'. One individual will have a slight advantage over another if it can produce more of the smaller gametes, as this will increase its likelihood of finding a partner. Because it has a limited budget of resources, gamete numbers will increase only if each gamete is smaller. Proto-males become locked into competition to produce smaller and smaller gametes, as the more gametes a proto-male produces, the more offspring it has, meaning this trait is passed on to the next generation.

In this process proto-males are not only competing but also 'cheating' the proto-females. The relationship is set up to be antagonistic from the start. As the proto-males gain an advantage over their brethren by making smaller gametes, a selective pressure is created whereby proto-females with larger gametes are more likely to successfully reproduce. With their diminishing size, proto-male gametes are no longer bringing much to the table when it comes to nutrients for the zygote. This means that if zygotes are to survive, the larger gametes must invest more and fill this gap. In their 1972 paper Parker, Baker and Smith describe males as 'dependent on females and propagate at their expense, rather as in a parasite-host relationship'. Flattering. Hanna Kokko tells me this view is often described as 'primordial sexual conflict'.

A constraint of unicellular life is that one individual can only create four gametes through meiosis. To provision an egg with enough resources to survive often means dividing into one large viable egg and three puny cells that are fated to die. Whereas four 'cheaper' sperm can be created by one meiotic division. Multicellularity greatly increases the number of eggs an individual can make, but the maths often shakes out the same. In humans it takes one meiotic division to make one egg, but four sperm. So sperm creators are once again 'cheating' by creating more reproductive opportunities for themselves.

This tale of competitive males and cheated females has been the dominant origin story for the sexes for decades, and the one I learned as gospel throughout my education and reading. So my mind was blown when I read evolutionary biologist Joan Roughgarden's refutation of this view of life in her 2009 book *The Genial Gene: Deconstructing Darwinian selfishness.* Roughgarden views it as a selection simply favouring an outcome that resulted in increased contact of gametes that could be anisogamy viable: 'The original PBS model is not about gametic conflict at all, it's about gametic contact.'[18] Essentially, the model describes a numbers game and the dramatic and gendered interpretation of it was just that, an interpretation, and a misguided one.

Roughgarden's own model – the RI model, developed with former student Priya Iver – takes conflict out of the equation, literally, as this intellectual fight takes place using mathematics. It presupposes that the ancestral isogamous species had tiny gametes, pointing to the alga *Chlamydomonas* as an example of a single-celled species with small gametes. In circumstances where small zygotes would no longer be viable, perhaps because they are the starting point for a larger multicellular organism, selection would favour individuals that could also produce large gametes. Sperm are maintained in the system and 'survival is maximised when one gamete is nearly the desired zygote size to begin with and the other gamete is tiny'.[19]

The RI model could be viewed as a 'male first' view, but in 2022 a 'female first' hypothesis appeared in the literature like an egg released into the primordial ocean. This hypothesis was developed by entomologist and evolutionary biologist

Yukio Yasui of Kagawa University and co-authored with his colleague Eisuke Hasegawa.[20] Whereas Roughgarden imagined our isogamous ancestor as possessing tiny sperm-like gametes like the single-celled *Chlamydomonas*, Yasui posits that their gametes were large and egg-like.

How did these ancestral species producing large gametes come to be? It might have been a transitional state early in multicellularity. Yasui argues that a multicellular organism provides greater resource-storage capacity than a unicellular organism. In this new, resource-abundant multicellular environment, selection would have favoured greater investment in gametes, creating 'inflated isogamy', where all members of the species now had large gametes. Males came later.

Yasui's paper explains that 'Inflated isogamy would have triggered one parent (becoming male) to cheat its mate (becoming female) by producing many small gametes'. This 'cheating' does not carry the same cost for the female as it did under the PBS model, as the females would not have to adjust their investment, which was already high. As smaller gametes increase the fertilisation rate, and the assumed benefits of sex, 'male cheating could provide a large benefit to females, and a smooth transition from isogamy to anisogamy would therefore be achieved'. While Yasui's paper uses the language of cheating, the logic is similar to Roughgarden's, emphasising mutual benefit as the driver of evolutionary change.

Does this clash of scientists mean we are left with a binary decision? Must we decide whether males and females arose from conflict or contact? The science of sex is seldom binary. This is clear in Kokko's work, which suggests that both competition and collaboration could have been active forces in the journey from isogamy to anisogamy. Kokko and colleagues have thought deeply about what conditions were like for our isogamous, ocean-dwelling ancestors. Were they living in dense groups? Or were they spread far apart? In their models this turned out to be a key question.

When organisms are in low-density environments, a female has an increased risk of her gametes never being found. 'Therefore,' says Kokko, 'it's actually quite nice that the other side invests in these tiny searchers.' So, in low densities, anisogamy can be viewed as a division of labour and collaboration between the sexes, females investing in resources and the presence of males upping the chance that the large investment will be fertilised.

The competition is on when the organisms are closely packed. In this situation the females would benefit if the males invested more in their gametes, but the density of males means that males who produce more tiny sperm continue to have the advantage. Kokko laughs as she explains this: 'That's actually quite funny to think about, whether you think this male or female thing is ultimately all about conflict or about cooperation [is due to] the density of long-dead marine organisms.'

The thinkers who have developed new hypotheses on the origin of males and females disagree about their findings, but they all have a common desire to challenge norms and suggest big ideas. Yasui described this as his 'think different' and tells me he does not like to 'obey authorities'. I interviewed Yasui over email as his Parkinson's disease made video conferencing difficult. Throughout his career he had been dreaming up various ideas about the evolution of sex but had not planned to publish them.

When he became ill, a friend encouraged him to go for it, and he has been fighting to get his big ideas out ever since, describing them as the 'fruit of my hidden lifework'. Despite the scale of his hypothesis, it was very hard for him to get published and he told me his paper was rejected by the prestigious journals *Nature*, *Science* and *Evolution* and he sensed pushback from the field. Despite this, the paper was featured in *Springer Nature*'s highlights for 2022, after being published in a smaller journal.[21]

The tendency to not obey scientific authorities is just as strong in Roughgarden. She received serious pushback for the RI model and other related proposals. Having read many of her critics, I believe some of the charges levelled against her were ad hominem attacks based on her gender. However, many came from a place of scientific integrity and respectful disagreement. Her view is still not accepted by a plurality of researchers today, but her challenge to the idea that conflict is at the centre of what it is to be male or female has left its mark.

For sexual conflict to exist between the sexes, one sex must incur a cost to their future reproductive potential that the other sex does not incur. Partly inspired by the PBS model and the Bateman–Trivers paradigm, the field of 'sexual conflict' has burgeoned in recent decades. As it turns out, even in fruit flies – a species where one scientist can investigate generation after generation in tightly controlled settings – determining whether conflict is shaping change can be surprisingly tricky.

Science has had an ongoing love affair with fruit fly sex. I found out just how sticky and contentious this affair can be talking to evolutionary biologist and fly ejaculate expert Ben Hopkins of the University of California, Davis.

In the 1980s it was shown that after female fruit flies had sex for the first time, their longevity was reduced. Good sex education advice for a female *Drosophila* would be the classic line from the teen movie *Mean Girls*: 'Don't have sex, because you will get pregnant and die!' The damage to the females was not directly caused by males during overzealous mating or courting. Part of the reason could be the investment the female made in producing eggs, but that could not explain all of it. What was going on?

The answer came in 2004, and it was hiding within the seminal fluid of male fruit flies. Their ejaculate contains a 'complicated molecular cocktail along with the sperm', Hopkins tells me. This cocktail includes lipids, carbohydrates, water, and genetic material. In fruit flies it also includes hundreds of different types of proteins, which might sound like a lot but is fairly restrained compared to the thousands of protein varieties found in human semen. The 2004 study by Stuart Wigby and Tracey Chapman found that a single protein in this cocktail of fruit fly semen, sex peptide, was largely responsible for the reduction in a female's lifespan and overall reproductive potential.[22]

The discovery of this chemical agent shown to be driving sexual conflict was big news, and the research community was jazzed. The idea was that sex peptide was an 'agent of manipulation' males used on females. The peptide interacts with a neuron in the female's reproductive system and small amounts enter her circulatory system as well. Hopkins says, 'There are broadly two states a female fruit fly can be, the un-mated state and the reproductive state, it's like flipping a switch.' It makes the female less interested in mating for several days, she will choose different foods, her gut activity changes, egg production increases, patterns of aggression change and she even sleeps at different times of day.

Female fruit flies mate with multiple males, storing sperm in their bodies. Researchers thought this might be a way males ensured paternity, by reducing the chance other males would add their sperm to the mix. 'Sex peptide became seen as a poster child for this field of sexual conflict,' Hopkins says. But was this case as cut and dried as it appeared?

Hopkins and his collaborator, Jen Perry of St Francis Xavier University, Nova Scotia, started to worry that the field might be getting carried away and overstating the role of conflict in the evolution of sex peptide. Hopkins' detailed understating of fly semen and its implications gave him the expertise to realise that some scientists were reading the literature 'selectively'. He emphasises that the problem was not with the original research, which he describes as 'relatively cautious with how it framed its results', but with the mass of work that followed and entrenched the little protein's role as 'the poster child' for sexual conflict.

If sex peptide was so bad for females, why did they have a receptor for it in their reproductive tracts? The location of this receptor seemed 'a little bit suspicious' to Hopkins.

He compared the life of a fruit fly in the lab to that of a wild fly. Life in the lab is like life in an all-inclusive holiday resort – flies have a continuous food source and are insulated from risk. However, he explains, 'Some of the estimates of how long *Drosophila melanogaster* live in the wild are five to seven days, whereas a lot of these studies keep females alive for 40, 50 days.' In this case, what is conflict in benign environments swaps to collaboration in harsh ones. When life is dangerous, it makes sense to kick the reproductive system into overdrive as soon as fertilisation might have occurred.

Hopkins believes that females may have evolved to use sex peptide as an efficient way for them to recognise mating has occurred. He also suspects it may have evolved as an 'elegant system' that, among other things, allows the female to track how much sperm she has in storage. He collated these arguments and more with co-author Perry in a 2022 paper entitled 'The evolution of sex peptide: sexual conflict, cooperation, and coevolution', which challenged the idea that sexual conflict was the sole force at work in this system.[23]

The role sexual conflict plays in driving evolutionary change may be overestimated in a range of species. This includes examples that to human eyes seem more violent and explicit than sex peptide. Male *Latrodectus* spiders, the genus that includes black widows and redback spiders, will tear into the exoskeleton of an immature female to mate. As she is not yet fully developed, her genitals are concealed, so this is the only way a male can inseminate her. She will then store his sperm, using it once she is reproductively mature. Gruesome, and a good candidate for sexual conflict. But when researchers checked to see if it negatively impacted a female spider's longevity or fecundity, they found it did not.[24]

Talking to Hopkins, I am fascinated by why he thinks sex peptide became a 'poster child' for sexual conflict. 'Battle of the sexes is a fantastically evocative framing, the research on it lends itself to quite exciting descriptions,' he muses. He emphasises that he is 'not a sceptic of the field as a whole' and points to another species of insect, the water strider, as a compelling example of sexual conflict. Male water striders have evolved elaborate grasping structures they use to force matings, and females have coevolved defensive structures that allow them to resist unwanted males.

Does 'primordial sexual conflict' lurk in modern marriages?

Love, sex and relationships can be hard, and full of pitfalls. For people interested in pursuing heterosexual dalliances, the emerging field of evolutionary psychology is offering a range of explanations as to how and why fault lines may appear.

Despite the difficulty of understanding sexual conflict in relatively controlled laboratory settings, many researchers have tried to apply the concept to human populations. This has involved using the ideas sketched by the Bateman–Trivers paradigm and the PBS model. Sexual conflict both lionises and denigrates men. On one hand, males are seen as active, vital competitors with agency and drive to do whatever it takes to perpetuate their genes. But on the other hand, they are cast as parasites, who are 'dependent on females and propagate at their expense', as Parker, Baker and Smith so proactively put it.[25]

Many researchers trying to explain the reproductive behaviours in the incredibly social, complicated, and diverse species that is *Homo sapiens* seem to be the most careless and enthusiastic when it comes to unthinkingly applying the Bateman–Trivers model. Trivers himself got the ball rolling in his seminal paper, musing on the reasons fathers may abandon their families:

> In the human species, for example, a copulation costing the male virtually nothing may trigger a nine-month investment by the female that is not trivial, followed, if she wishes, by a fifteen-year investment in the offspring that is considerable. Although the male may often contribute parental care during this period, he need not necessarily do so … Given the initial imbalance in investment the male may maximize his chances of leaving surviving offspring by copulating and abandoning many females …[26]

Ouch. It is not easy to read but in some ways it 'feels true'. Trivers' application of parental investment theory sets up an inherent conflict between men's and women's sexual strategies. Men gain more if they up sticks and spread their wild oats. Women gain more by finding a 'good man' to settle and invest in her and the kids.

The arguments embedded in the research Trivers popularised, which I repeat were based on an error-filled fruit fly study, were picked up by science and the popular imagination. For instance, in his 1998 book *Why is Sex Fun?* respected science writer Jarred Diamond muses:

> A man, for example, produces 200 million sperm in one ejaculate – or at least a few tens of millions ... By ejaculating once every 28 days during his recent partner's 280-day pregnancy – a frequency of ejaculation easily within the reach of most men – he would broadcast enough sperm to fertilise every one of the world's approximately 2 billion reproductively mature women ... That's the evolutionary logic that induces so many men immediately after impregnating her and to move on to the next woman.

Is it, Jarred, really? The logistical mastery it would take to achieve such a feat is impossible. Selection for complex behavioural traits, such as serial impregnation and abandonment – something rare enough that I think Diamond is giving men short shrift here – happens in specific contexts over generations, a context which has never included the impregnation of 2 billion women by one dude.

Maybe I am being too harsh on Diamond. He is exaggerating, sure, to make his point and it is not something a reader is expected to take seriously. But this logic has also made its way into the scientific literature, where it is more worrying. In a 2003 paper, psychologist and founder of the International Sexuality Description Project, David Schmitt, writes: 'Whether one woman mates with 100 men, or is monogamously bonded with only one man, she will still tend to produce only one child in a given year.' True enough. He continues, 'A man who is monogamous will tend to have only one child with his partner during that same time period.' True again.

But his logic comes crashing down as he writes:

> One man can produce as many as 100 offspring by indiscriminately mating with 100 women in a given year … this presents a strong selective pressure – and potent adaptive problem – for men's mating strategies to favour at least some desire for sexual variety.

If a healthy guy is good at seducing woman, why can't this busy Lothario impregnate 100 women in a year? In her book *Testosterone Rex,* Cordelia Fine has kindly done the maths on this. She sets our hero up to win. His sperm is not depleted, he is not in competition with other males and every woman he sleeps with is fertile. Considering the chance for two people with no fertility issues of conceiving after sex is around 3 per cent, she estimates the chance of a man conceiving 100 babies in a year as 0.00000000000000000000000000000000000000000000000 0000000000000000000000000000000000000363.[27]

He could up his odds a bit by sleeping multiple times with each woman. But still, the guy is dreaming. Fine's lucky fella had perfect conditions in which to succeed, but as she points out, men's reproductive lives have not been easy. For most of human history they have lived in small communities. Many potential partners would have been too old, too young, too related, too pregnant, breastfeeding or otherwise infertile.

Science is the pursuit of the truth – or at least claims that can be backed by evidence. Evolutionary selection took place in real human communities over tens of thousands of years, not in a psychologist's fantasy scenario. What makes this so frustrating is that Schmitt's research was an interesting and extensive cross-cultural survey of men and women and their preferences for sexual partners. The data showed a tendency for men to seek greater sexual variety than women. While the data is interesting, it is hard to trust the analysis that surrounds it when it is based on such a loose understanding of basic biology.

For male philandering to be selected for in the way Trivers suggested, over history some men would have monopolised reproduction, out-breeding men who stayed true to one woman.

Instead of turning to fantasy to understand the reproductive lives of men and women, psychologist Gillian Brown of the University of St Andrews and colleagues have turned to data.[28] Long-term population data recording maternity and paternity in populations is hard to come by, and always arrives with some caveats, as without genetic testing, paternity – and in some cases maternity – can be hard to ascertain. However, the researchers were able to get robust records from 18 societies around the world – some dating back to the 1700s. The data covered various marriage systems, including monogamy, serial monogamy, and systems where one man can marry many women, termed polygynous. While there are some societies where women can have multiple husbands, there was not enough robust data from those societies for this study.

According to Brown's research, whether reproductive success varied more between men than it did between women depended on the culture being examined. When all the data was averaged, male reproductive success was more variable, but it was by no means the rule.

The paper warned against assuming that a society where one man can have multiple wives will necessarily mean that a small number of men are able to monopolise the reproductive opportunities. While this was the case for the Dogon of Mali, it was not the case for the Aka of the Central African Republic, where reproductive success was close to even between the sexes. They point out that in many polygynous societies, the practice itself is relatively rare, with fewer then 5 per cent of men having many wives, and that many of these cultures allow for divorce and remarriage, so women can have multiple husbands over their reproductive life.

Brown and colleagues also show that in some societies multiple wives may add to the wealth of a household, meaning the kids of a man with many missuses are well provisioned for. In other contexts, multiple wives can reduce the wealth of a household, making it hard to raise kids and potentially limiting men's reproductive success. The prevalence of sexually transmitted diseases could also impact how having multiple wives allowed individual men to increase their reproductive success compared to other men.

In western countries such as the United States where serial monogamy is common, men's reproductive success varied more than women's. This has been put down to men remarrying younger women, however, in other serially monogamous cultures such as the Ache of Paraguay and Pimbwe of Tanzania, whose societies are smaller in scale, both men and women have children with multiple partners.

The takeaway? Brown and colleagues write: 'Little strong evidence is currently available to plot the shape of Bateman curves for men and women across human populations.' They call for more research into the area but emphasise this work 'throws into question Bateman's expectation of universal human sex roles'. Humans are incredible in our flexibility, our capacity to live in diverse situations and to allow culture to shape us. This means that over our history, the power variable of reproductive success as a driving force in evolution may have ebbed, flowed, and reversed through time and space. Making universal declarations about how the Bateman–Trivers paradigm has impacted our sexual and romantic yearnings suspect.

Despite this, I have been consistently shocked by how many researchers who study sex roles in humans apply the Bateman–Trivers paradigm paired with the idea of sexual conflict to explain differences between men and women. In his book *Why Men Behave Badly*, evolutionary psychologist David Buss describes the subconscious motivation for intimate partner violence as a strategy to reduce their mate's 'value' and thereby stop other men pursuing her. This logic even seeps into more tangential research. A 2021 paper invoked Bateman and sexual conflict to provide an evolutionary explanation of the different leadership styles taken by male and female politicians during the Covid-19 pandemic.[29]

The stakes of rolling these ideas out before there is enough data to support them are high. People are desperate for information that will help them navigate relationships. Emphasising a worldview steeped in conflict without exploring other dynamics profoundly impacts the way people see themselves and each other. I was scrolling through the comments on a YouTube video where podcaster Joe Rogan interviews evolutionary psychologist and mate-choice expert David Buss.

User @skanda1832 wrote:

> When I learned about this subject matter while studying Biology in college, it permanently tainted my ability to fully embrace romance and relationships, having seen how fundamentally primal and opportunistic the game is.[30]

This sentiment was shared by many other commenters. Buss has used the Bateman–Trivers paradigm to test a raft of hypotheses about male and female, conflict, behaviour, and preferences. Much of his work is compelling, and I am sure some of it will stand the test of time. But it is built on shaky assumptions about how selection has operated in our ancestral past.

Hold up! Is there even a male–female binary?

Anisogamy can lead to the idea that when it comes to sex, male and female are neat binary categories. But nature is far beyond the binary in nearly every way you slice and dice it. Just as DNA has taken four bases and used it to code for the multiplicity of life on Earth, sex has taken a general tendency for gametes to come in two different sizes and spun it into a delightful assortment of sexual systems, expressions, and strategies.

I got in touch with wildlife biologist Elissa Cameron of the University of Canterbury to understand how sex informs science in the field. She has worked with meerkats in African deserts, marsupials in Tasmanian forests and horses in the mountains of New Zealand. She explains we tend to sort things into male and female because 'that's nice and easy, but it's not that simple, actually'.

A major complication of anisogamy is that it is a single term that encapsulates a huge variety of systems. For example, both the Northern Pacific sea star and the kiwi bird are anisogamous. A female Northern Pacific sea star can produce over 20 million tiny eggs, which she will broadcast into the water column to be met by smaller sperm.[31] In contrast the flightless, ground-dwelling kiwi of New Zealand produce a single abdomen-crushing egg that can be more than 20 per cent of the female's body weight.[32]

Many insect species have shockingly large sperm. A fruit fly called *Drosophila bifurca* currently holds the record for the world's largest sperm, with a gamete that reaches nearly 6 centimetres in length when fully unfurled, although it is usually tightly packed within its body. Whereas large mammals, like elephants, tend to have tiny sperm.[33] The diversity captured by the term anisogamy leads Cameron to reflect that it is 'mind blowing' that we think it is meaningful to divide all the animals in the world into two categories 'when we actually defined [the sexes] really randomly'.

Knowing the gamete size of an animal, without further detail, is not enough information to make foolproof inferences about how that animal reproduces or behaves. While evolutionary forces may select for owners of one gamete size to be pushed in a particular direction, such as caring for young more than the other sex, it is in no way predictive. A classic example is the seahorse, where females lay eggs into a male's pouch. He will then go through a pregnancy, his belly swelling like a human parent, before giving birth to tiny baby seahorses. The pregnancy is complex.[34] The pouch provides energy and calcium for the growing embryos and removes wastes – essential functions for all live-bearing animals. Many of the 3000 genes a seahorse uses to perform these tasks are analogous to the genes expressed during vertebrate female pregnancies, including in mammals.

Many species have distinct 'morphs' of a sex, where members of the one sex will look and behave very differently to each other. An example of this is the ruff, a wading bird that is named for the elaborate neck feathers of some breeding males. Most males fight for territory in a shared mating ground, and display to females through dance. Their neck feathers can be a rich range of earthy colours. Other males do not fight for territory and these birds always have a white ruff. The rarest males look very similar to females; they hang out with the females and were unknown to science until 2006 because they were assumed to be female.[35]

These morphs are genetically determined, so are inherited and not a result of age or other factors. This is an amazing, unusual, and exciting system to evolve. Instead of celebrating this cool system, scientific and popular discussion of this newly discovered smaller, more feminine male morph can be derisive. In 2015 one headline in the *Washington Post* declared: 'A "supergene" turns these male birds into female impersonators or sneaky mate thieves – for life.'[36]

The sneering tone used to describe males who subvert the dominant view of masculinity gives us a window into the cultural challenges we face when trying to understand how sex is operating in nature.

All in the chromosomes? How do you become female or male?

A dip into any reef around the world means diving into a melee of sexual diversity. Many species on the reef are simultaneous hermaphrodites, producing two sizes of gametes. This includes many of the corals themselves, as well as animals such as flatworms, barnacles, and even some fish. While some species of fish can be male and female at the same time, many can flip from one sex to the other and sometimes even back again. This sex-switching superpower is not limited to the ocean. Duck owners have reported that female ducks will transition to male ducks if their ovaries are injured, even producing functional sperm.[37] Sex swaps needn't be the result of things going wrong. Green frogs have been found to switch sex at rates of 2 to 16 per cent. This was originally thought to be a result of pollution, and met with distress. But later research showed it happened in unpolluted areas, so is likely just a regular part of life.[38]

My understanding of sex was forever changed when I was reporting for ABC Science on new research by molecular biologist Erica Todd from the University of Otago into what causes blue-headed wrasse to change their sex.[39] These fish can change from female to male extremely quickly, disintegrating their ovaries and developing testes in just ten days. They also undergo a makeover. Females of the species are small, with a yellow back and a white belly. Males are larger, with a chunky blue head and a yellowish-green body. The species lives in harems of one male and a group of females. If the male is removed, this will trigger one of the females to start her transition. Within minutes, her behaviour changes, she begins to defend the territory of the group and starts showing sexual interest in the females.

I had always had the unexamined assumption that even if sex could change, there would always be some sort of fundamental genetic difference between males and females. In reporting on the blue-headed wrasse, I learned that was not true. Fish do not have sex chromosomes – you could not tell a male and female blue-headed wrasse apart just by looking at what was in their genes. When they become male, genes that 'feminise' the fish are progressively turned off, and the genes that 'masculinise' it are turned on.

Todd told me that sex is not determined by chromosomes but by some kind of cue. For the wrasse, that cue was the male disappearing and triggering the stress response in the female.[40] That stress response then opened the floodgates to the genetic cascade that made the fish male. Temperature is another common cue that determines sex, especially in reptiles, many of which can change sex as embryos.

The phenomenon is less common in mammals, but humans can change their secondary sexual characteristics – those traits generally associated with one sex that are not directly related to reproduction, such as the ability to grow facial hair or having breasts – by taking hormones. In these cases the hormone production is outsourced, and the cue that sparks the change is a decision by the individual taking them.

Mammals can appear to be the most binary of animals. For many mammals, unlike fish, the genes an individual inherits play a role in determining whether that individual will develop into a male or female. However, even in mammals, Cameron explains, sexual traits are more of a 'continuum … it is definitely not a binary, though it might have binary elements to it'. When sperm is being made through meiosis, half of those sperm will have XX chromosomes and half will have XY. On the Y chromosome is a gene called the Sex Determining Region gene, or SRY gene for short. The gene does not code for hormones that directly act on sex, rather it binds to other DNA, changing its shape, and changing which genes are being expressed, leading to the formation of characteristics such as testes. 'It's not the X or Y itself, it's what's coded on it,' Cameron emphasises.

The road from genes on a chromosome to sex being expressed is long and winding. 'It should be simple, but then it's just not,' Cameron tells me. You might have the chromosomes and the genes to be male or female, but to fully develop down one track, 'you've got to get the right pulses of the right hormones at the right developmental stages, [and] that has to continue up until sexual maturity [and beyond],' she says. When this does not happen, sexes can turn up with characteristics not typical of the types of chromosomes they have. Differences within the sexes are not always genetic. 'You get some cool effects, if you're a female rodent raised in utero next to male siblings, you become relatively more "masculinised",' she explains. 'And if you're a male raised next to two female siblings in utero, you become relatively what we would call "feminised".'

Crucially, what it is to be male is not lying in the Y chromosome, just as what it is to be female does not live on the second X chromosome. In fact, the genes that encode for sex in humans are scattered around the genome, and it does not take much to shake things up and change the way sex will develop. For instance, androgen insensitivity syndrome will often lead to people with XY chromosomes developing what we think of as female genitalia. They will have internal testes and secondary sex characteristics at the extreme 'feminine' end of the spectrum and very little body hair. People with male XX syndrome are born with two X chromosomes and no Y chromosome. This is a rare condition affecting only one in 20 000 people. Some people with XX male syndrome have ambiguous genitalia and others have what is thought of as 'standard' male genitalia.

While most mammals have an XX–XY system, it is not universal. A group of rodents called mole voles certainly do not play by the chromosomal rules. Some mole voles have an XY male XX female system, but in other species males and females have XX chromosomes, and in still other species males and females only have one X chromosome.

Intersex is integral

Among all the sexual systems I learned about in my entire zoology major, I do not remember learning about intersex animals – individuals in a species that has two sexes who have characteristics 'not typical' of males or females.[41] So I was surprised when, in the 'methods' section of a paper on Tasmanian devils, I saw that an intersex animal had been caught.[42] The animal was not included in the subsequent data or mentioned again, as the research was looking at differences between males and females. This made me wonder whether intersex animals were routinely removed from datasets. Were scientists unknowingly erasing them from our understanding of the natural world?

I started asking around. Many scientists I knew who studied marsupials – mammals that raise their young in pouches – had stories of intersex animals. A devil scientist and friend, Jarrah Dale, told me he had seen an intersex devil with a pouch, viable young and an external scrotum that looked as if it contained testes. He has also seen devils with penis, testes, and rudimentary pouches. Another researcher, Rodrigo Hamede, told me he had seen only two intersex individuals in 18 years of devil research. He speculates that they are rare, but there would have been other intersex individuals that he had encountered unknowingly because outwardly they looked either male or female.

Jumping into the literature I found that intersex individuals were not limited to marsupials – in fact they were found across the animal kingdom. While intersex conditions are found in low rates in most species, the sheer number of animals on Earth mean they are not a trivial addition to animal life.

Cameron, who is no stranger to encountering intersex animals in her research, believes that the scientific attitude to them is changing. She tells me that if an intersex individual was observed by a researcher '50 years ago, it would be seen as an aberrant individual, a genetic anomaly, a dead end … but how you treat them has changed dramatically and probably really quickly over the last few years'. This change has come as variation within a species, which could include variation in how sex is expressed, has been understood to be crucial to how that species operates within its environment.

This change in attitudes has not been universal. In his 2020 essay, 'In humans, sex is binary and immutable', Stanford University geneticist Georgi Marinov writes:

> Humans are also born with a great variety of devastating congenital deformities and diseases, and if alien exozoologists were to write a description of *Homo sapiens* based on extensive observations of the population, such a description would never feature, for example, anencephaly, and neither would it include anything else but binary sex.[43]

This is contrary to how Cameron, an earthly zoologist, goes about things. She explains to me that how she includes intersex individuals in her research depends on their density in the population and whether the questions she was trying to answer related to population averages or individual variation, emphasising that if they were observed, they would be recorded. But maybe aliens would have more old-school methods.

Marinov's paper continues that intersex individuals only occur from spontaneous mutations not inherited from a parent or recessive genes becoming unmasked – the same process that will result in two parents with brown eyes who carry a recessive mutation for blue eyes having a child with blue eyes – or disruption in embryonic development. He leans heavily into the 'dead end' view of these individuals, explaining that they often cannot have children, inferring that they should not be considered when trying to understand our evolutionary history. In this, I think Marinov misunderstands the way genetic diversity enters a population and how natural selection operates.

Most mutations have a negative impact on an individual's capacity to survive and reproduce. In very rare instances a mutation will be beneficial. These rare beneficial mutations are what has slowly, over time, led to the great diversity of life on Earth. Mutations that lead to intersex variations are no different – usually, but not universally, making it less likely an individual will reproduce.

Evolution needs diversity to operate. If there are no differences within a population, no specific traits can be selected for. In North America a small percentage of grizzly, black and polar bears are born with intersex conditions that mean the bears have sex and give birth through an enlarged clitoris that appears like a penis.[44] While this is considered an intersex trait in bears, it is standard for hyenas. All female hyenas give birth out of large clitorises that they also urinate out of. To me this opens questions about under what conditions specific intersex traits could be selected for and over time become the new 'standard' way a sex within a species exists.

This tendency to view animals that do not fit the binary as 'aberrant' or a 'dead end' means that discussion of intersex animals, when it does happen, often pathologises them. But as Tom Levy of Stanford University and colleagues point out in a 2020 paper about intersex Australian red-claw crayfish (*Cherax quadricarinatus*), 'The common occurrence of natural intersexuality could also suggest a possible fitness advantage.'[45] In other words, intersex individuals may provide advantages at a population level.

They then put this idea to the test. A crayfish, or lobster, can be intersex in many ways, just like a human or a bear. But a common way is to have male sex chromosomes but egg-producing reproductive equipment. Levy and colleagues found that after an intersex female's eggs were fertilised, three-quarters of their young were female, compared to the standard 50/50 male/female split. This would result in a higher ratio of females in the population, meaning the population can grow faster, as more females mean more eggs to fertilise. These intersex females are rare, so when populations are large, they do not noticeably throw off the population's sex ratio. The researchers suggest that when colonising new territory, or responding to a dramatic reduction in population, these intersex females might give the species the edge they need to survive. While they pointed out that this was early days in the research, they are some of the first scientists to apply their curiosity to the positives of intersex animals in wild systems.

What is the future of the sexes?

In recent years our deepening understanding of the diverse ways sex-linked traits show up in the natural world has led to zoologists calling to expand our binary view of sex to better understand the natural world.[46] This call has also been heard in the medical field, where scientists and medical doctors are finding that a strict binary limits their capacity to understand the human body and deliver appropriate care.[47] Since the mid-20th century, many people with intersex conditions have undergone medical interventions as infants to make them fit the binary – a practice associated with serious medical side effects. While advocating for bodily autonomy, physical autonomy and self-determination, intersex organisations have also been critical of binary views of sex.[48]

The call to push our scientific understanding of sex outside of the binary has received pushback. In their 2023 paper 'Biological sex is binary, even though there is a rainbow of sex roles', Wolfgang Goymann of the Max Plank Institute and colleagues emphasise that sex comes down to one thing and one thing only: gamete size.[49] For them, there are two sexes and that is the end of the story. They point out that factors often used to describe sex such as chromosomes, genitals, sex roles, hormones or social constructs do not define the sexes, so variation within these traits does not challenge the binary.

Does this mean we are left with another binary decision: sex is either binary or it is not? Cameron reflects that 'the only actual definition [of sex] that stacks up is the size of your gametes'. But for her this is not the end of the story. 'Western culture is very much about putting things into boxes and breaking them down into the smallest possible units, whereas Indigenous knowledge systems are often inclusive and talk about interlinking and overlapping.' To Cameron, observing animals in wild systems, these interlinking and overlapping understandings of sex may have more utility.

Following these debates, I feel I am stuck between a couple arguing. Each is convinced they are right; each has evidence to bring to the table, but ultimately they have different understandings of what they are fighting about. Primatologist Agustín Fuentes of Princeton University put it best when he wrote: 'The bottom line is that while animal gametes can be described as binary (of two distinct kinds), the physiological systems, behaviours and individuals that produce them are not.'[50]

Those who seek an expanded definition of sex are often talking about chromosomes, genitals, sex roles, hormones, and – in the case of humans – gender roles and identities. While these can be described as traits associated with sex, and therefore not part of sex themselves, they are often integral to how sex is experienced and understood, both in humans and in the non-human world. A binary gametic system can be accepted, but the science suffers when that system is used to imply a binary outcome.

CHAPTER 4

Why do we make love?

Sex for a garden sparrow is a blink-and-you'll-miss-it affair. Two birds flutter together, cloacas touch, sperm is transferred. Wham bam thank you ma'am. Sex is perfunctory and impersonal for many species. A lot of ocean-dwelling critters do not even have to touch, they just spray eggs and sperm into the water. For other creatures, sex is a drawn-out affair. It is not uncommon to find two beetles joined by the genitals for hours as they go about their lives. Many animals' sex lives are confined to periods when conception is possible. Humans are different. We often have sex when we are not fertile, whether that be due to pregnancy, older age, time of the month, the sex of the partners, or myriad other reasons, often finding great pleasure in it. Our sex lives often come with a messy side effect: love.

A little note on language before we dive in. Sexual behaviour can be tricky to define, but, unless otherwise specified, when I write 'sex' in this chapter I am talking about behaviours that can, but don't necessarily, facilitate sexual reproduction, or behaviours closely tied to those reproductive behaviours. The English language has really let us down here.

Science has sped ahead in the last hundred or so years, taking our understanding of the universe to places we could never have imagined. But if some science is a sleek, electric-powered racing car speeding down the highway, research into non-reproductive sex and love is a beat-up old bomb of a car.

The driver is nervous, the handbrake is stuck and the journey includes several checkpoints with suspicious and moralistic guards who insist on looking under the hood for something suspect.

Why? A good deal of our modern reticence about studying sex and love can be traced back to a bloke born in medieval Sicily in 1225: the enormously influential philosopher and theologian, Saint Thomas Aquinas. While he died centuries ago, his ideas on sex and marriage have survived. These ideas were spread around the world as European powers colonised huge chunks of the Earth and are still referenced in law. Despite the odds, researchers have found a way to do fantastic and brave work on the evolution of pleasurable sex and love. But to understand the unique constraints this field still experiences, it is helpful to know why and how it was supressed for so long.

Old texts might be musty and dusty, but humans have always been lusty. I knew that as a priest and friar Aquinas had taken a vow of celibacy and thus had committed to not having sex, so I assumed his writing would be stodgy and detached. Although I disagree with his perspectives on most things, I was surprised to find a kinship in his work – he was wrestling with the same kinds of questions I am. Namely, what is the role of sexual pleasure? Is it to create a romantic bond between people? How do we live with each other well? He approached these questions with genuine interest, observed the natural world and considered how his observations applied to humans.

Aquinas's views on sex were part of his broader philosophical outlook called Natural Law.

He argued that humans and other animals behave in ways that are instilled in us by a divine creator: we desire to stay alive, reproduce and provide for our offspring. In this view of life, animals cannot act immorally, as everything they do is according to this plan. But we humans have a pesky thing called free will. It is our free will that can lead us to act out of step with nature, and thus into sin. Aquinas argued this was especially true sexually. But, as humans are also rational, we can use our faculties to divine how to act under Natural Law.

To get the tick of approval from Aquinas, sex must involve sperm entering the vagina, no contraceptives, and be done in a context where any children conceived could be cared for. For him, sex outside these circumstances was wrong. This includes things most modern people still consider taboo, such as bestiality, rape, adultery, and incest. But it also encompassed types of sex that have a wider, if not universal, acceptance, like masturbation, sex with someone you are not married to, and homosexual sex.

Aquinas classified sex outside his parameters as not 'natural'. But if God has made us and imbued us with our desires, how can anything we do not be natural? Aquinas believed sexual pleasure, in moderation within marriage, was God-given to promote procreation. Under Natural Law, the purpose of sex is procreation, although there is a little loophole for married couples who cannot have children, recognising the role sex can have strengthening a relationship. Importantly, God's plan for us is not just for us to have children, it is also to raise and educate them.[1] Sex outside of marriage was against Natural Law because the children created were not raised by both parents. He points to the role of solicitous bird dads at the nest, arguing that married fathers are part of Natural Law.

If you can provide for your kids as a single mum, or for that matter in a throuple or in a commune, is sex for fun OK? Surprisingly, Aquinas did consider the case of rich single mums.[2] He did not give them a pass, however. For Aquinas, moral behaviour was determined by what affected 'the whole species, and not by those things that accidentally befall one individual'.[3] So, what is best for most, is best for you. The same logic holds for unmarried couples.

This is a book on evolution! Why am I talking about religious teachings from the 13th century? I am not here to critique religion. My point is that this philosophy, which is baked into cultures with a Christian tradition, can corrupt our capacity to faithfully observe and interpret the natural world, and it still does today.

Natural Law really struck a chord in Christian cultures. When Catholicism started to fragment during the 16th century and Protestant churches popped up, this world view continued. It was a profound part of the education and legal systems of the scientists who founded evolutionary biology, and still is for many biologists today. Since Aquinas, a common way to talk about or legislate against gay sex is to call it an 'unnatural act' or an 'act against nature'. In my state of Tasmania, acts 'against the order of nature' are still in the criminal code and until 1997 these included any sexual acts between two men.

It cannot be underestimated how much the 'sex is only for reproduction' dogma worked its way into our cultural understanding of what sex is, could be and should be. While many evolutionary biologists are firmly secular, the Aquinas view of sex runs deep and at first glance seems to mirror evolution by natural selection. Because, as evolutionary biologist – and famous atheist – Richard Dawkins put it, 'life has no higher purpose than to perpetuate the survival of DNA'.

It is true, sex facilitates reproduction and the perpetuation of shuffled genetic codes into the next generation. Sexual behaviour and reproduction are so closely linked in our minds, that it can make it hard for scientists to conceive of how the behaviour might be operating in ways other than just copulation and reproduction. But researchers who have widened their perspective have found the history of pleasure, sex, connection, and love is far more fascinating than a simple collision of gametes.

Want to know about non-reproductive sex? Ask the experts!

Same-sex sexuality is often seen as a spanner in the works of an evolutionary worldview. By its very nature it is non-reproductive, therefore it forces us to understand why sex might exist outside the context of procreation. For this reason, pondering same-sex sexuality is a great way to begin understanding how humans evolved such a diverse sexual repertoire.

As I write, most scientific papers will tell you that 1500 species have been recorded engaging in same-sex sexual behaviour. That number was first quoted in Bruce Bagemihl's paradigm-shifting 1999 book *Biological Exuberance: Animal homosexuality and natural diversity.* Considering that book is now 25 years old, it is likely that hundreds, if not thousands of species have joined the parade.

Same-sex sexuality is abundant, but despite this the scientific community systematically kept nature in the closet for centuries. Science journalist and radio producer Lulu Miller created what she called a taxonomy of suppression, which outlines three common ways this information is kept out of the scientific literature:[4]

(1) If you expect all sex to be straight, gay sex will be misclassified as straight sex, especially in animals where males and females look similar.

(2) You see it, but you do not record it. This might be because of a scientist's homophobia. An example is when

mammologist Valerius Geist first encountered male mountain goats going at it. He later told a journalist, 'I still cringe at the memory of seeing old D-ram mount S-ram repeatedly. To conceive of those magnificent beasts as "queers". Oh God!' He eventually changed his perspective, admitting to suppressing the information, and said that the rams live in 'essentially a homosexual society'.[5]

A scientist might also self-suppress because of fear of reprisal. In times and places where homophobia is rampant, reporting on same-sex behaviour in the animal world can bring suspicion on yourself. This is not just a problem of the past, but a real concern for many scientists today.[6]

(3) Overt suppression. You saw it, you recognised it, and now it's in a paper ready to go. But no one will publish, or funding is threatened.

Eventually, due to the brave work of many scientists, the prevalence of same-sex sexuality became impossible to deny. Despite this, reviewing the research on it will give you the impression that same-sex sexuality is odd, strange and a little bit queer. Sentiments such as this from a 2013 paper are common:

> Although male–male mounting is well documented in many ruminant species, as goats, sheep, or cattle, it is inconsistent with the basic law of nature: procreation.[7]

The spectre of Aquinas remains present in the language here, framing same-sex sexuality as 'inconsistent with the basic law of nature'.

Ecologist Julia Monk from the University of California, Berkeley, tells me that the way same-sex sexuality was talked about in the literature 'not only felt alienating to me as a queer person but also, from my perspective as a queer person, it seemed counterintuitive'. When she started digging into the research, she noticed an implicit assumption that same-sex sexual behaviour had evolved independently in each of the 1500-plus cases: 'There's this constant framing of "How could this crazy, *ab*normal behaviour possibly arise?"' This seemed unlikely to her, so she decided to flip the script and ask, 'What if a diversity of sexual behaviours is the ancestral condition for sexually reproducing animals?'

Monk started with the idea that the first sexually reproducing animals would have produced both sperm and eggs. For these animals there is no such thing as same-sex sexuality. Every adult is a potential mate. Monk contends that the earliest animals with different sexes would have pursued mating regardless of the sex of their partner. Why? Discerning who is male and who is female is not an easy task. If we assume that as soon as the male and female sexes separated within a species, exclusive heterosexuality became the norm, we also have to assume that these ancient creatures were equipped with the capacity to easily distinguish who was packing which gametes.

To tell the egg-makers from the sperm-sprayers, females and males must develop differences to signal which sex they are. They must also develop the skill to interpret those signals. This kind of separation and discrimination 'should come later in the evolutionary process', Monk tells me. Placing energy into this differentiation and distinction could have been costly, slowing down the process of sex. In this context, when it is hard to tell who's who, indiscriminate mating might be the most advantageous way to find a partner. Monk's logic is, 'If there were not high costs to that behaviour and if there was still sufficient reproduction happening, I saw no reason why such a behaviour would be eliminated … It seemed to me like there were not so many mental stretches if you began from that perspective to explain the prevalence of same-sex behaviour today.'

The logic is neat, but does it play out in nature? There are still animals that have sex much like the early broadcast spawners. These are the echinoderms, which include the sea stars, sea urchins, sea cucumbers, and sand dollars. They have sex via the 'spray and pray' method, releasing eggs and sperm into the water column and hoping they will connect. But to up the odds of fertilisation, they will often clump together before doing this. Monk found that for many species the 'predictor of if they would release their gametes was the degree of overlap with another individual, not necessarily the sex of that other individual'. This method is close to the ancestral state she hypothesised. Some species, like the sand sea star (*Archaster angulatus*) are a bit more discerning. Generally, males will crawl onto a female's back before they release sperm. But male–male pairs are common. Natural selection does not imply evolution of perfectly optimised systems; often it results in evolution of the good enough.

Monk's hypothesis is not based on her own lab or fieldwork. 'What we presented was an idea that we think is backed up by evidence from the broad realm of literature on this topic,' she says. Her thinking is an invitation to the field to go out, do the research and hopefully back it up.[8]

This challenge was taken up by Brian Lerch as he was doing his PhD at the University of North Carolina. Impressed by Monk's logic, he wanted to see if evidence would support the hypothesis.

For Lerch, the compelling idea within Monk's research was that same-sex sexual behaviour was being maintained in species because it was adaptive – that is, it was selected for because it increased the chance of survival and reproduction. In his work Lerch uses mathematical models to understand biological systems, so he developed a model that would test Monk's idea. When his models included a 'cost' for only attempting opposite-sex mating, 'imperfect sexual discrimination' would evolve. The numbers backed Monk up.[9]

While Lerch's maths supports Monk, he is careful to clarify that it does not apply to species with complicated social lives. He thinks about these evolutionary ideas in terms of echinoderms and fruit flies, explaining his work is built 'on assumptions that I think just don't apply to social organisms'.

In social species, like our closest relatives the chimpanzees and bonobos, there are strong hypothesises about the evolution of same-sex behaviour. Most of this thinking is underpinned by the idea that it increases social bonds, group cohesion and diffuses potential aggression. Monk is careful to point out that her work does not contradict any of these ideas. Her challenge to researchers is not to assume that the behaviour 'arose out of nothing ... It could have been already present as some of the natural variation'.

We will get to how sex is operating in these species. But aren't we missing why most people want to have sex? Because it feels good!

Uncovering the roots of pleasure

The clitoris is often described as the only organ in the human body whose sole function is pleasure. Throughout the history of science, male animals, male humans, and male genitalia have all received more attention than their female counterparts.[10] As most clitorises are owned by females, this has resulted in the clitoris being hard to find – for scientists in the 20th century, at least. Even once located, uncovering its evolutionary origins and the origins of its orgasmic powers has proved challenging.

The most famous part of the clitoris is the glans. This small external tip of the clitoris is packed with sensory receptors. In humans, it hangs out above the vaginal opening, at the prow of the vulva just south of where the vaginal lips meet. The glans is often mistaken for the whole clitoris but is just the tip of the iceberg. Most of the organ is inside the body. Bulb-like structures extend down from behind the glans and almost hug the vaginal canal. The organ is nerve rich, and responsible for sexual pleasure.

The most sensitive part of the of the clitoris is the glans, and for many people stimulation of it is how they achieve orgasm. The structures hugging the vaginal canal can be stimulated through penetration, meaning orgasms achieved through penetration are, in fact, clitoral, although there is some evidence to suggest that stimulation of other parts of the vagina can also be involved.[11] For the sake of clarity, when I discuss clitoral orgasm it will be referring to an orgasm reached through stimulation of the clit via external stimulation, internal stimulation, or a combo of the two. This is to distinguish it from orgasms achieved through the penis, prostate, nipples, mind, or anything else.[12]

Whether a person's genitalia will develop into a penis and testes, a vagina and clitoris, or a combination of the two is determined by hormones in utero, but both are built from the same original structures, so share many similarities. For instance, much of the clitoris has erectile tissue, and just like a penis, blood will flow to the area and enlarge it when sexually aroused.

Medical training about the clitoris in the 20th century was abysmal. It was often omitted from curriculums or, when acknowledged, the basic anatomy being taught was inaccurate. This continued into the 2000s. Thankfully this is starting to change, due in large part to the work of Helen O'Connell. She was Australia's first female urologist and is still a working physician in Melbourne. As a student she found her textbooks included page after page of detailed descriptions of the penis, and very little on the clit. Meaning, if you walked into a doctor's surgery with a penis, the doctor would know a lot more about what to do with it than a clitoris – scary.

Motivated to improve matters for the clitoris, O'Connell set about mapping it. After dissecting ten cadavers aged between 22 and 88 at the time of death, she had a detailed picture of the organ, the tissues it was made of, and its nerves. This work was published in 1998.[13] When she compared her findings to the textbooks current at the time, she discovered they included many inaccuracies, such as describing the bulbs of the clitoris as being associated with the vaginal wall or the labia, rather than being part of the clitoris.[14]

Frustratingly, O'Connell was reinventing the wheel. The anatomy of the clitoris was described in detail by anatomists in the 19th century. This included astoundingly accurate illustrations by Georg Kobelt published in 1844. Even back then, Kobelt was frustrated at the lack of attention paid to the clitoris and vagina.

When did medicine lose the clitoris? In the 19th century science was split when it came to sexual pleasure and masturbation across males, females and intersex people. Sexual pleasure was valued as part of health by some, and seen as suspect and in urgent need of suppression by others. By the early 20th century, things became more binary and an emphasis on women finding sexual pleasure through penetration put the clit, especially the glans, into disrepute. There was even

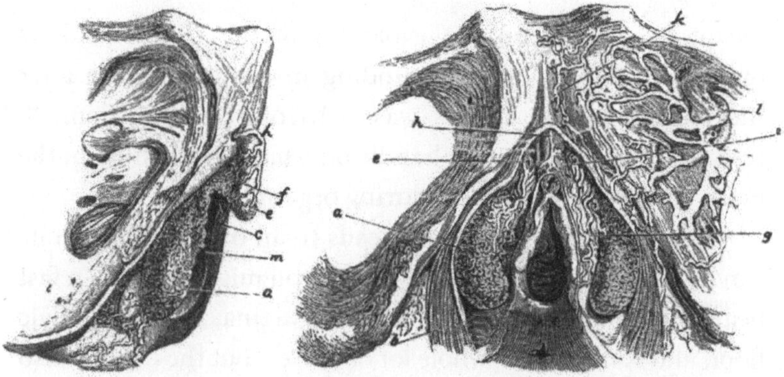

Lateral view of erectile structures of external organs in female (left). Blood vessels were injected, and skin and mucous membrane were removed. *a*, bulbus vestibule. *c*, plexus of veins named pars intermedia. *e*, glans clitoridis. *f*, clitoral body. *h*, dorsal vein of clitoris. *l*, right crus clitoridis. *m*, vestibule. *n*, right gland of Bartholin. Front view of erectile structures of external organs in female (right). *b*, sphincter vaginae muscles. *e*, venous plexus of pars intermedia. f, glans clitoridis. *g*, connecting veins. *k*, veins passing beneath pubes. *l*, obturator vein.

Dissection of the pubic region with clitoris.
Preparation of Georg Ludwig Kobelt [1844]

a misguided movement to increase women's sexual pleasure during penetration by surgically removing the glans.

This attitude seemed to leak into medical education. In a 2005 paper, O'Connell further describes the organ with the use of micro-dissections and magnetic imaging and notes that the:

> … study of anatomical textbooks across the 20th century revealed that details from genital diagrams presented early in the century were subsequently omitted from later texts.

She goes on to say that the evidence suggests this was the 'result of active deletion rather than simple omission in the interests of brevity'.[15] That's science talk for 'there was a clit cover up'!

This suppression was medically dangerous and denied people an understanding of their own bodies, or the bodies of their lovers. But it also had another effect. How were biologists meant to understand the evolution of the clitoris and its orgasms without a basic grounding in anatomy? Soon after the biology of the clitoris was rediscovered by O'Connell, researchers started to get a handle on what is happening in the brain and hormonal system during orgasm.

When clitoral stimulation leads to an orgasm, the results can be dramatic. There are the things you might expect – a fast heart rate, muscle contractions in the vagina, anus and pelvic floor, and sometimes a whole lot of noise – but the orgasm also reaches the brain, with brain imaging showing brain activity across much of the organ steadily building to a maximum at the moment of orgasm then decreasing again.[16] And, of course, it is usually incredibly pleasurable – although even in consensual sex some people have an orgasm without the accompanying pleasure.[17]

Hormones too join the party. Blood rushes to the pituitary gland, a tiny structure at the base of the brain that makes and releases hormones when another part of the brain called the hypothalamus tells it to.[18] Among these hormones is oxytocin, which in social animals is associated with reduced anxiety, enhanced social memory, and the release of dopamine, which is part of the brain's reward system – it tells you 'well done'. Hormones called prolactin and luteinising hormones are also released, which are involved in the reproductive cycle.

How did this elaborate response evolve? For the penis the answer seems clear, at least at first blush: orgasm results in ejaculation, and sperm needs to be released for reproduction. But a clitoral orgasm is not necessary for pregnancy, so there had to be more to it. Even without a full understanding of the anatomy, some researchers had a crack at figuring it out.

One set of ideas suggests that while the orgasm might not be necessary for conception, it can up the odds enough that it has been selected for over time. In his 1967 book *The Naked Ape*, Desmond Morris suggested that the exhaustion and sheer sexual satisfaction of an orgasm kept a woman lying horizontal after receiving ejaculate, thus making fertilisation more likely.[19] The disturbingly named 'up suck' hypothesis was very trendy in the early 2000s and contended that muscle contractions during orgasm help suck sperm into the uterus. This has now been disproved in a series of studies.[20] Some studies suggest having an orgasm, or at least having sex, after artificial insemination can increase the chance of pregnancy, but not be necessary for it. Another idea is that the pleasure of orgasm incentivises sex, so our ancestors who had bodies that enjoyed sex had more of it, and thus more offspring, and over time there was a slow, steady building of the capacity for sexual pleasure.

These ideas all strongly link the orgasm's history to reproductive sex. For Elisabeth Lloyd, a distinguished professor at Indiana University, this did not necessarily have to be the case. She has argued strongly that no evolutionary pressure led to this type of orgasm, and it is just a 'happy accident' caused by the shared developmental origins of the clitoris and the penis. All these hypotheses were developed before our deepened understanding of clitoral anatomy and orgasm from the early 2000s. They also contain a baked-in assumption that orgasms evolved in early humans, or at least in primates.

In 2016 biologists Mihaela Pavličev and Günter Wagner launched a new hypothesis, made possible by an increased understanding of the human clitoral orgasm, that considered whether the origins of these explosive events might be older than first assumed. Pavličev is not a sex scientist. The evolutionary biologist mainly studies viruses, placentas, and pregnancy, and tells me she did not set out to develop a theory on the origin of the orgasm, but came to this research via an unexpected path. She first started to consider the orgasm question while studying preterm birth in humans. To tackle this, she needed a deep understanding of the reproductive cycle. This involved comparing reproductive biology across species, and it was this process of comparison that led to her big idea.[21]

Ovulation is when an egg, or ovum, is released from an ovary. This is the beginning of its journey to the uterus, where fertilisation can occur. In humans, this cycle is pretty much set, generally occurring around every 28 days from puberty onwards until menopause. Breastfeeding and pregnancy disrupt it, so people who have many pregnancies can have relatively few cycles in their lives. Our cycle is set by a sort of biological calendar. But this is not the case across all mammals. Species as diverse as the rabbit, the alpaca and the koala need to have penetrative sex to ovulate.

Our system is called spontaneous ovulation, because it happens spontaneously irrespective of sexual activity. Species where sex is needed to get things going are called copulation-induced ovulators or male-induced ovulators. While some species are on one end or the other of this spectrum, some fall in between, with sex increasing the chance of ovulation or ovulation only being possible from sex at certain times of year. Going through a reproductive cycle is energy intensive – for most humans who do it, it downright sucks – so only ovulating after sex is a great energy-saving trick for species that may not come across a mate all that often. It also reduces the chance of meeting said mate at the wrong part of a cycle and not being able to reproduce.

For copulation-induced ovulators, the act of sex signals to the brain that sex has occurred. The brain passes the good news on to the pituitary gland, which gets busy pumping out hormones, such as luteinising hormone. These hormones make their merry way to the ovaries to signal ovulation. But how does the brain know to get this ovulation party started? Is it chemicals in the semen? Does it just *know*? Or does stimulation of the clitoris and possible orgasm tell the brain it's go time? Looking across species, Pavličev noticed the same hormones were common triggers for ovulation across mammals – ours too is triggered by luteinising hormone. But strangely, these same hormones are released after clitoral orgasm in humans.

The hormonal similarities between human orgasm and ovulation made Pavličev wonder. Was the orgasm, or a precursor to it, originally a trigger for copulation-induced ovulation? To prove that our lineage has lost copulation-induced ovulation, copulation-induced ovulation must have predated spontaneous ovulation. This can be done by comparing species across the tree of life. How?

Let's use humans and tails as an example. Humans have no tails but monkeys do. What came first, tails or no tails? As we know how different primates – monkeys, our fellow apes, and curious creatures like bush babies and lemurs – are related to each other, we can figure this out. We can create a primate family tree. The trunk of the tree is the common ancestor all primates share. When the tree branches, that represents lineages diverging, which can happen again and again. Just as how, in a human family tree, the end of a branch is shared by siblings, with cousins on a separate branch, all sharing an ancestor further back in time.

Looking at the primate tree of life, our closest relatives, the apes, have no tails. Monkeys are our next closest relatives after apes, and they do have tails. As do lemurs, who branched off the family tree before the monkey–ape split. From that we can infer that the common ancestor of all apes had a tail, and we became tail-free sometime after our split with our more distant shared monkey ancestor. Tails came first, then were lost in our lineage. This means having no tail is more recent, or in evolution jargon, it is a derived trait.

But there is a spanner in the works. Europe's only monkey species, the Barbary macaque, has no tail. Does this make our hypothesis about apes wrong? No. This suggests that the Barbary macaque evolved this trait independently of apes.

Pavličev did the same thing with copulation-induced ovulation and spontaneous ovulation. She found that copulation-induced ovulation did indeed evolve first, and spontaneous ovulation came later, independently cropping up in species as diverse as mice, dolphins and dogs.

Pavličev's paper focused on mammals, but the clitoris seems to have been present before mammals evolved. It shows up in reptiles and in the embryos of birds, although it seems most birds lose it as they develop in the egg, suggesting the origin of the clitoris probably dates to the evolution of penetrative sex in vertebrates. In a 2022 paper on female sexual anatomy, famed animal vagina expert and advocate for more clitoris research Patricia Brennan notes, 'Many other fundamental questions in the study of female genitalia remain unexplored.'[22] She encourages more comparative studies between the genital anatomy of different species and for research on female genitals to be normalised. Brennan must have been chuffed later that year when the news broke that snakes, previously not known to have clitorises, turned out to have two per female – and who doesn't love a twofer?

But how to test whether an orgasm, or at least clitoral stimulation, occurred in copulation-induced ovulators? Antidepressants such as Prozac make it famously difficult to come. So Pavličev administered Prozac to rabbits, let them have sex, and checked whether they ovulated. The idea was that if the drug interfered with orgasm, that might impact ovulation. And it did! Just as some people can still climax on antidepressants, ovulation was not completely suppressed in rabbits, but it was reduced by about 30 per cent. To make sure the Prozac was blocking the bunny's orgasm, not inhibiting ovulation more directly, the researchers also gave some dosed-up bunnies a hormone that initiates ovulation. In these cases all rabbits ovulated equally.[23]

Not all researchers were convinced by Pavličev's rabbit-orgasm evidence. An article covering the research in *Scientific American* contained challenges from two scientists on the grounds that a rabbit's orgasm and a human's cannot be compared.[24] Raúl Paredes at the National Autonomous University of Mexico was concerned that pleasure is at the heart of an orgasm and contended we cannot know whether other animals feel sexual pleasure. Julie Bakker, a ferret ovulation expert at the University of Liège, argued: 'There's no such thing as orgasm in rabbits … it is more like a light switch, in which male stimulation triggers the brain, which triggers ovulation.'

I am not sure about these arguments. While measuring pleasure in other species might be tricky, it does not follow that they do not experience it. Additionally, the hypothesis does not rely on pleasure being present; neural and hormonal response may have been the origin of the orgasm, with pleasure coming later.

We do not just ovulate differently to rabbits. We also have very different clitorises. In humans there is quite a distance between the vaginal opening and the glans, meaning that penetrative sex is not always a sure-fire way to induce orgasm. But in many other species, such as the rabbit, the glans is inside the vagina, meaning penetrative sex always has the potential to stimulate it and presumably makes climax through penetration easier to achieve. What is going on with us? Why are our clitorises playing hard to get?

Pavličev returned to her evolutionary tree of spontaneous ovulators and copulation-induced ovulators. She had a list of which animals had the glans outside, inside or on the border of their vaginal canal. While it was not a one-to-one relationship, the glans of species who were copulation-induced ovulators tended to have the most sensitive bit of the glans inside the vagina, ensuring it was stimulated by penetration. This included rabbits, hedgehogs, and koalas. Species with external glans seemed to be spontaneous ovulators. This included all primates she had information for, a bat, and some rodents such as rats, mice, and guinea pigs. Species that had the glans on the border (maybe a best of both worlds scenario?) were a hodgepodge of spontaneous and copulation-induced ovulators. Goats, sheep, dolphins, anteaters and dugongs are all spontaneous ovulators with the glans on the borderlands, whereas armadillos, alpacas and ferrets all needed sex to ovulate. To Pavličev this supported the idea that orgasms evolved as a mechanism to induce ovulation, since when copulation was needed for ovulation, the animal's biology dictated stimulation was inevitable.

Pavličev's hypothesis might answer why the orgasm evolved, but it does not explain why it has stuck around after spontaneous ovulation removed its original function. Did it gain new functions? Unfortunately, Pavličev was not able to research this, and it wasn't for lack of trying. After submitting a funding application to research possible links between orgasm and the immune system she was denied funds. Not only that – 'They asked us not to apply again,' she says. What made this strange is that the paper was a hit outside academia. The idea made headlines worldwide. It 'was everywhere, my parents knew about the paper before I told them', Pavličev tells me. Her parents saw the story in 'a local paper in their small Slovenian town'. But despite the fanfare, and the letters from women thanking her for her work, she did not get a good reception in the medical world.

Shocked by this, I ask why, when sexual pleasure is so important to human health, many in medical research were so dismissive.

'That is what they didn't agree with, that it is important for human health,' she replies. Members of her research community told her it was 'irrelevant'. Their view was that if clitoral orgasms are not connected to disease, then they do not need to be studied in medicine. 'I think it's a limited way of thinking because, if you don't know what's healthy, then it's hard to understand what's disease,' Pavličev says.

She stresses not all roadblocks to her research were due to colleagues not seeing its value, and many did. But as research bodies are often linked to conservative funders, there was a nervousness around supporting the work in case other funding was threatened. 'They don't want to have to choose between being able to treat kids or support research like [orgasm research]. Of course, there's no question, one has to agree with that,' she says.

Despite the lacklustre response from the medical world, the scientist believes her hypothesis is a valuable part of our evolutionary story: 'One can sell this as the emancipation of female sexuality, in the sense that not every penetration causes pregnancy.' Pondering how an external clitoris may have shaped evolution, Pavličev suggests it could have 'changed the relationship between the sexes and could have selected for patient, more empathic males. Because it just takes a while.'

I would go further. The movement of the glans liberates sexual pleasure from the necessity of penetration – greatly expanding the kind of sexual acts that can give pleasure. This is certainly the case for humans. Research suggests those trying to achieve clitoral orgasm through heterosexual penetrative sex reach orgasm less often than people trying to achieve it alone or with a same-sex partner, two types of sex that would be a lot more difficult with an internal clitoris.[25]

But what of the penis? In a comprehensive refutation of Pavličev's hypothesis, psychologist and orgasm researcher Barry Komisaruk pointed out:

> ... ejaculation can occur as a spinal reflex, as reported in men with severed spinal cord, so it is not correct to assert that ejaculation requires orgasm. Hence, it is likely that orgasm plays important roles in males (e.g. reinforcement of sexual behaviour) other than just ejaculation and sperm transfer, and perhaps the behavioural reinforcing effect was at least as significant an evolutionary pressure as ejaculation, an evolutionary reproductive pressure that could apply equally to women.[26]

Compared to vaginas and clitorises, penises have received a huge amount of scientific attention. But the evolution of pleasure and the penis has also been, from my perspective, somewhat overlooked. The 'mystery' of orgasms derived from the clitoris is due to their apparent disconnect from reproduction. For males, it seems clear cut: an orgasm is needed for ejaculation that is needed for pregnancy, and it feels good because that incentivises sex. It's a good story, and probably true. But, just as clitoral orgasms are a complicated event involving the brain, hormones, blood flow and muscle contractions – so are orgasms achieved via stimulation of the penis. Perhaps it is time for the penis to gain some of the mystery given to the clitoris?

However it came to be, humans are blessed with genitalia that can provide pleasure outside of copulation. This means our sexual behaviour can, and does, get put to other uses. Perusing Pavličev's list of animals that have easily accessible glans and spontaneous ovulation, a pattern occurs. Many of them, such as primates and dolphins, are famous for having a lot of non-reproductive sex and complicated social lives. Is that sex as pleasurable for them as it can be for us?

Discussion of pleasure is often missing from animal research, but this is starting to change. In 2022 evolutionary biologist Patricia Brennan and colleagues published an in-depth description of the dolphin's clitoris.[27] This research found that the dolphin females' sexual organs were sizeable and had features uncannily similar to human clits. They contain erectile tissue and chunky nerve bundles, with a riot of nerve endings under skin that was three times thinner than the surrounding skin. Can they come from them? In an interview with *Cell Press*, Brennan answered this, saying, 'For their sake, I hope they do ... but we actually can't really answer that question, because studying the sexual response in dolphins is very hard.'[28]

The structure of the organs suggests to Brennan that they are fully 'functional' and do give these marine mammals pleasure. This is backed up by circumstantial evidence. Dolphins, like humans and many primates, will have hetero sex all year round, even when they are not able to conceive. Same-sex sex is also common. Female bottlenose dolphins are known to stimulate each other's clitorises with those bottlenoses, as well as with their fins and tails. Brennan hopes that by comparing the biology of dolphin clitorises with other species, researchers will be able to further map the history of clitoral sexual pleasure.

Dolphins of both sexes also masturbate using objects. Much to many parents' chagrin, masturbation is also common in primates at the zoo. A grey seal will use its flippers to rub its penis, and goats have been observed engaged in autofellatio.[29] This commitment to genital stimulation suggests to me it is likely these species derive pleasure from it.

Sex and social relationships

Once your sexual pleasure has been liberated from the chains of reproduction, what can you do with it? In the dense forests beside the Congo River in the Democratic Republic of Congo lives a species that makes enthusiastic use of its external genitalia: the bonobo. We are equally related to bonobos and chimpanzees, our closest cousins, having shared an ancestor with them around 5 to 9 million years ago.[30] In the 1990s they became a bit of a sensation when it was revealed just how much lesbian sex the species engaged in. Female bonobos, like humans, have a glans that is external to their vagina. They will rub their genitals together in a move called genito-genital rubbing, or GG rubbing. As they rub they often look into each other's eyes, sometimes reaching what appears to be orgasm.[31]

It's not just the ladies getting down. In this species sex occurs between males, between males and females, and across most age classes. Most of the sexual behaviour bonobos engage in is social, not resulting in pregnancy. For bonobos, sex is part of a rich social life, as it is for many primates. Relationships within the species are close, and they live in 'fission-fusion' communities where groups will separate, rejoin, and change over time. Females in this species form close relationships and have a lot of social power, often monopolising food and ganging up on males.

It seems that sex is the secret sauce keeping these communities cooperative. Bonobos will often have sex to reduce tension between individuals. Why fight when you can make love, right? They will also have sex after conflict, and this is known as appeasement sex. Multiple heterosexual matings in a period of fertility can also help keep the peace in a more sinister way – a 'confusion of paternity' is thought to reduce the risk of infanticide, even in species where it is unlikely individuals understand the link between sex and paternity. Bonobos will also have sex when there is nothing to compete over and nobody has fought, as a form of social bonding. Cute.

The species even has specific gestures and vocalisations used to solicit sex. For example, a pant hoot, which is described as a 'low-pitched, melodious call sounding like a whining "hoo-hoo"' can be used to solicit sex from one female to another. If a male pant hoots at a female, it is an invitation to reproductive sex. A 'scream' is an invitation for appeasement sex for both males and females.[32]

In humans, sex for pleasure often brings people together, deepening bonds of connection. The same appears true for female bonobos. This was shown in a study where a group of wild bonobos habituated to scientists were observed for 1483 hours. The resulting paper written by Liza Moscovice and colleagues explains that '971 independent sexual events' were recorded.[33] After the bonobos had sex, researchers would take samples of their urine and analyse them for the hormone oxytocin. From studies in other species, we know this hormone is linked to the formation of emotional bonds and associated with a range of cooperative and coordinated behaviours. These range from human mother–infant bonding, grooming in chimpanzees, bonds between monogamous partners in humans and monogamous primates, and even the relationship between dogs and their humans.

The researchers were keen to find out if bonobo sex spiked oxytocin. It did – but primarily after GG rubbing and not after opposite-sex intercourse. Why? The researchers are not certain but conjecture it could be because sex between females involves eye contact, which is associated with oxytocin release in other species. Eye contact is less common in opposite-sex sex, which did sometimes cause these spikes, but not enough to show up in the averages. Orgasm spikes oxytocin in humans, so the researchers suggest the spikes might mean GG rubbing is more likely to induce orgasm for females compared to when they had sex with males. The degree the oxytocin spiked after sex was variable, which again is consistent with humans, where more intense orgasms correlate with higher levels of the hormone.

After a bout of GG rubbing, females were more likely to stay in proximity to each other compared to opposite-sex couples after sex. Females who had sex also cooperated more, teaming up with each other to confront or chase off other individuals. This led the researchers to ponder if positive interactions between pairs of females that have sex a lot are more common, as their relationship is closer as they have, as the scientists put it, 'more opportunities to engage in emotionally rewarding sexual interactions'. In other words, lots of good sex leads to good relationships for bonobos. To me this study is significant as it draws a strong link between sex, social connection and cooperation, underpinned by physiological evidence.

What does all this jungle loving have to do with humans? Despite being closely related, we are very different to bonobos. But we share an unusual trait: the ability to closely collaborate with unrelated individuals who aren't our mates. Other species that do this are chimpanzees and some bottlenose dolphins. In these species evidence is starting to come in that sexual behaviour is one of the ways in which individuals of the same sex bond.

For bonobos, sex clearly has a role in developing the close relationships needed to live in a functioning, generous community. Neuroscientist and animal brain expert Andrew Barron of Macquarie University made this connection between prosociality and sex in humans and bonobos in a paper he co-authored, arguing that in recent evolutionary history there has been a 'strong driver' for 'reduced reactive aggression, increased social affiliation, social communication, and ease of social integration'.[34] This has led to the rise of a whole suite of traits: 'In many prosocial mammals, sex has adopted new social functions in contexts of social bonding, social reinforcement, appeasement, and play.'

Bonobo and dolphin behaviour says to me it's possible that our human desire to 'make love', to have sex that is not only pleasurable but connects us emotionally, got its start as our species was becoming more peaceful and cooperative. Crucially, this could be the case for all types of sex, not just sex that leads to reproduction. As Barron writes, 'We argue that for humans the social functions and benefits of sex apply to same-sex sexual behaviour as well as heterosexual behaviour.'

What about love?

Romance: wonderful, painful, joyful and the spark of great art and literature. People love thinking about love, so I was surprised to find science is rather quiet on the subject. When it comes to love, it seems scientists get a bit embarrassed.

So who is going there? I got in touch with Adam Bode, a PhD student at the Australian National University who describes himself as the 'only person in the world [solely] studying the [scientific] evolution of romantic love'. Although he emphasises that there are some researchers who look at the phenomenon outside an evolutionary context, they are few and far between. Romance was only identified as a legitimate field of study in the 1970s, and was especially popularised by two psychologists working in the US: Ellen Berscheid and Elaine Hatfield. While their work was instrumental in starting the field, it was also a warning, as they had their funding publicly stripped after a senator from Wisconsin claimed it was a 'waste of taxpayer money'.[35]

It was not only the money the senator objected to, it was the notion of putting a scientific lens on love. As he put it in a press release:

> I'm also against it because I don't want the answer …
> No one can argue that falling in love is a science. The impact of love is a very subjective, nonquantifiable subject matter. Love is simply a mystery.[36]

A romantic notion, but I don't think the spark felt by two lovers as they hold each other close is made less wonderful by understanding the hormones coursing through their bodies or the evolutionary story containing the millions of lovers who brought them to that moment. Just as understanding the physics and chemistry of a galaxy does not make a night sky less breathtaking.

This aversion to love in the sciences has perhaps resulted in evolutionary science being clinical and detached as it tries to understand human connections. Bode explains that there are a small number of evolutionary psychology labs that study sex, and each one has a strong focus on reproduction. While conceding that 'ultimately the mechanism through which evolution occurs is reproduction', he is nevertheless critical of this approach and points out that the questions being asked often take love out of the equation. Researchers 'usually talk about mating', and a lot of evolutionary psychology focuses on 'the decision to select a mate either for short-term or long-term mating'. While also considering himself a human-mating scientist, Bode is explicitly adding the question of romantic love into the mix.

This is Bode's definition of romantic love:

> Cognitive activity of romantic love includes intrusive thinking or preoccupation with the partner, idealisation of the other in the relationship, and desire to know the other and to be known. Emotional activity includes attraction to the other, especially sexual attraction, negative feelings when things go awry, longing for reciprocity, desire for complete union, and physiological arousal. Behavioural activity includes actions toward determining the other's feelings, studying the other person, service to the other, and maintaining physical closeness.[37]

This usually describes those first few heady weeks, months or years of a relationship, or the desire for one, but in some cases this state can exist for decades, although many people slip into companionate love, which he describes as 'that stable friendly love that long-term married couples have for each other'.

Humans often form strong bonds between two individuals that may produce offspring. None of our great ape cousins – the gorillas, orangutans, bonobos and chimpanzees – form long-term reproductive pairs, which is called pair bonding. Bode thinks it is a crucial ingredient in romantic love. Considering pair bonding is a pattern we do not share with our close relatives, Bode suggests it may have evolved after our ancestors parted ways with the lineage that would eventually become chimps and bonobos. Bode's investigation of romantic love has led him to believe its evolutionary roots can be found in the bond between mother and offspring.

In mammals, who nurse their young on milk, the mother-offspring bond is often intense, especially while the young are, well, young. The relationship goes both ways and human babies usually start forming strong preferential bonds by six months old. Mother and infant experience bonding attraction – that is, they are motivated to stay close to each other and are distressed when parted. They also think about each other obsessively, or at least a lot.

Hormones such as oxytocin and vasopressin, a hormone associated with protectiveness and self-defence, or defence of others or territory, are highly associated with this bond. In research that has primarily been done on cisgender mothers who birthed their children, when human mums are shown images of their babies, predictable areas of the brain light up, including the left ventral tegmental area (VTA), which is associated with reward, learning and addiction.

For many new mums the experience is incredibly like romantic love – at least in terms of body chemistry. Bode believes these similarities suggest that psychological underpinnings of romantic love may have been 'co-opted' from the existing pathways involved in the mother–offspring bond. The co-option of an existing trait into the formation of a new trait is a common way evolutionary change occurs, sometimes facilitating dramatic transformations.[38] We saw this in the example of the glans clitoris, which potentially began as a mechanism to induce ovulation and now gives pleasure and has a role in social connection. Bode's hypothesis contends that this pre-existing hardware to facilitate mother–infant bonds was tweaked slightly to facilitate romantic love. He further supports this idea by pointing to the prairie vole, a small rodent with a relatively simple social life that forms pair bonds. Evidence from this well-studied monogamous species also suggested that their tendency to pair bond was indeed co-opted from the mother–infant bond.

Love, war and chimpanzees

A female and male pair of chimpanzees sometimes sneak away from the community for a multiday private sexcapade, or consortship, while a female is in oestrus and able to conceive. But these liaisons do not translate into long, close social bonds. And it is the pair bond that seems to be at the heart of romance. Until 2023 it was thought that humans were the only great ape to experience pair bonds, but to chimpanzee researcher Aaron Sandel of the University of Texas, this didn't seem quite right. Sandel spends hours of his life traipsing around a field site called Ngogo within Kibale National Park, Uganda. His study subjects are mainly young male chimpanzees. Male chimpanzees can have close relationships that last for years, and Sandel now believes they are pair bonds, just without the childcare.

It was Sandel's PhD supervisor, John Mitani, who first quantified enduring relationships between male chimpanzees. He did this by analysing data from ten years' worth of field studies showing that males formed equitable, reciprocal social relationships that could last for years. This research came out against a backdrop of studies suggesting pro-social behaviour between males was only ever in the context of acquiring short-term gain.[39] Sandel has taken this a step further, arguing these relationships are pair bonds: 'Male–male relationships are the closest thing to romantic love among chimpanzees.'

Bonded males will often travel together, sit next to each other, and preferentially groom each other. Sandel tells me these relationships can be short lived but some last 'five, six, seven, eight, nine, ten years, this isn't a trivial interaction, this is a real relationship'. As a field biologist, Sandel collects quantifiable data using a phone app to log the specific behaviours individuals display, so their lives can be subject to statistical analysis. With all the hours he spends with the animals, he gets to know them pretty well.

Two chimps he believes are current 'candidates for a pair bond' are a young adult male called Evans and a middle-aged male called Mulligan. Sandel tells me that Mulligan has 'a different way of being in the world as a chimp … he's much more a caretaker', who will often look after and groom young chimps and does not seem to engage in struggles for dominance. He and Evans are close, they often hang out and groom. Sandel remembers sitting and watching when the pair reunited after being separated for some time, maybe foraging in separate bits of forest. 'I don't know how long they had gone without seeing each other, but they got so excited, and they rubbed genitals face to face and one put his head in the other's groin.'

Male chimpanzees show a lot of sexual behaviour towards each other, so just because Mulligan and Evans seemed to be getting it on does not necessarily mean it was facilitating social bonding. As Sandel says, 'In animal behaviour it is so difficult to study emotions.' Sexual behaviour between male chimps occurs in a variety of contexts and Sandel suspects they can be broadly divided into two categories, the first being to facilitate a bond between individuals, and the second being a kind of reassurance.

Sexual behaviour often arises at time of high tension and when individuals are nervous around each other. Sandel observed the chimpanzees at Ngogo as one community split into two, and this polarisation was showing up as conflict between the groups. When males from opposing neighbourhoods met, they would often engage in sexual behaviour. Sandel reflects that it is 'almost like they're trying to keep a conflict at bay'.

To distinguish between these behaviours, Sandel is now trying to record behaviours known to be associated with positive or negative states during these interactions. This may clarify how behaviours like mounting map onto relationships and emotions. For Evans and Mulligan, it does seem that their sexual contact occurs in the context of a relationship that includes affection. Sandel says, 'They're not engaging in aggression between each other, but they do spend a lot of time together, they groom, and when they meet, they're excited and they mount.'

Eventually after a period of tension, full-blown conflict broke out between the neighbouring groups of chimpanzees Sandel was studying. For this species, violence between groups is conducted by males. Males patrol their territory together and cooperate to kill their enemies. Sandel was interested in pairs during conflict. He noted that not all males had a pair bond, and after five years of conflict many males had died, so had perhaps lost partners. He remembers one time 'after this battle, the first chimps to start patrolling back to confront the other faction, were pairs of friends ... In one case maternal brothers and in other cases these unrelated young-ish adult males.' Emphasising that this was interpretive, not a data-driven scientific hypothesis, he speculated whether 'they had that bravery because they were with their partner'.

Sandel is still hoping to discover how these bonds are formed, what roles they play and how they differ from other social bonds. But the foundation is there and in 2023 he published a hypothesis suggesting that same-sex pairs in our ancestors – like the ones chimpanzees enjoy – might have been co-opted to form the basis of human romantic love. Considering that strong social relationships also exist between female bonobos, Sandel reasons that in the 'search for the origins of romantic love, it's going to be in same-sex friendships of animals like our ape ancestor'.

How might this have occurred? Humans now have, as Sandel puts it, 'the whole gamut of relationships. We have intense friendships, we have romantic love, and they can manifest between all genders'. Looking at the close ties within genders of our closest evolutionary cousins, Sandel believes it might have been intense same-sex social bonds that were then co-oped to form the basis of romantic love, which often in humans involves an enduring connection between people who raise offspring together.

What about the mother–infant bond that Bode proposed? Both researchers acknowledge that exactly how the pair bond came about is not accounted for in Bode's model. Sandel thinks that there may be other species, such as the prairie vole, a species that forms intense reproductive pair bonds, where the bond was co-opted directly from the mother–infant relationship. But in primates he thinks there must have been an 'intermediate step' and believes same-sex friendships and bonds might have been that step, noting that friendships are less one-sided and more cooperative than mother–infant bonds, and can happen at a variety of life stages.

Bode thinks that Sandel's hypothesis could 'feasibly be right', with pair bonds evolving first and romantic love 'sitting over the top'. However, he also thinks this is unlikely, pointing to the strong likelihood that this was not the pattern for prairie voles. The species' penchant for long-term relationships was likely co-opted via the mother–infant bond, and there is no evidence of an intermediate step. He also suspects that romantic love in humans might be more intense than the pair bonds in the chimps Sandel studies. The shared obsessive nature of the mother–infant bond and romantic love suggest to him that there might not have been intermediate steps. Both researchers are only just beginning to test their hypotheses, so whether either will stand up to rigorous testing, and if they do, how they will complement or compete with each other, is yet to be seen.

Romance is a strong component of most cultures, suggesting that however it evolved, the capacity to fall in love has been strongly selected for. Bode believes this is because pairs who were bonded via romantic love were more likely to have children who survived. Mothers and children might have also been more likely to survive if male romantic partners were shielding mothers from potential violence and danger.

Human males are far more invested in their young compared to chimpanzees or bonobos, although caring souls like Mulligan show that even among chimpanzees, this trait is on a continuum. It is possible that the story of romantic love is wrapped up with the story of increased investment by fathers. Although the existence of aromantic people who do not experience romantic love shows that, like many human traits, there is diversity in the way our hearts work.

Shaking off the procreative doctrine

There is a sexual revolution underway in evolutionary biology. Researchers have begun to get more incisive understandings of sex, pleasure and love by imagining evolutionary stories that do not exclusively focus on sex in the context of reproduction. Many of the thinkers I found most exciting while researching this area were young and developing their hypotheses as PhD students, like Monk and Bode, or early in their careers, like Sandel. They are building on the brave work of people who risked ridicule or reprisal to highlight all the ways sex, or the anatomy of pleasure, show up in nature. This makes me hopeful that today's generation of scientists will continue to creatively explore exciting new ideas.

So, armed with these new ideas, how do scientists continue the revolution? To find out, I got in touch with Michelle Rodrigues of Marquette University in Wisconsin. Her work has a focus on evolution and social relationships in primates. A challenge she has identified is that there is disagreement in the field as to what is considered sexual. For example, many of the seemingly sexual interactions between males were originally only classified in categories like 'reassurance' – the sexuality got 'taken out of it'.

In many species, opposite-sex sexual behaviour gets classified as 'sexual', whereas same-sex sexual behaviour gets classified as 'social'. When hetero sex is lumped into the 'sexual' category, without further consideration, the social complexities of the behaviour can be lost. There is a third category, called sociosexual behaviour, which encompasses both and captures any sexual behaviour no matter the gender. Different research groups and communities that spring up around specific species do not agree what behaviours should be described as social, sexual or sociosexual. Rodrigues explains this makes understanding primate sexuality tricky: 'We don't actually have a lot of good comparisons across species to understand common primate behaviour versus species-specific behaviour versus social rituals that only emerge in certain social groups.'

I have argued that the narrow view of sex, only viewing it in the light of reproduction, is a hangover from the Christian teaching of Natural Law. On the other hand, evolutionary biology and primatology has a long tradition in non-Christian countries, most notably Japan. Rodrigues tells me that the methodology and research aims within Japanese primatology often differ from European and North American research groups. The Euro-American field focuses on 'primates as models for understanding human behaviour and understanding human evolution', whereas Japanese groups 'weren't looking at primates to be a model for understanding human evolution as much as trying to understand social complexity, and how individual dynamics and the different relationships form the social group'.

As evolutionary biology becomes more diverse, both within and between countries, different approaches, focuses and perspectives will likely add texture to the evolutionary story of sex, pleasure, and love. This will be more possible if the people with power in universities, funding bodies and research groups defend the right of scientists to investigate these questions and support diverse voices.

Saint Thomas Aquinas looked for moral guidance in the behaviour of other species as he understood them. I do not think this is a wise idea. Violence is common across species, but if I started a brawl, I do not think I could dodge blame by saying 'chimps do it'. However, scientists who have respectfully and curiously researched animal systems without preconceived ideas of what they ought to find have begun to uncover a world where other species engage in pleasure and connection in ways many people could never have imagined. While we do not need to look to these animals for guidance, I think we can appreciate that their experiences and evolutionary love stories echo our own.

CHAPTER 5

Why do we get cancer?

I was a soccer kid, always kicking a ball around the dusty school oval at lunch time. There were goals missed, spectacular saves and slow-forming friendships made chatting when play was at the other end of the oval. But the most vivid memory from my soccer days is when an announcement came over the school loudspeaker: 'Can Zoe Kean please come to the principal's office? Zoe Kean to the office.' I do not remember the looks on my teammates' faces, I just remember bolting the length of the school to get to that office. That morning my mother, Jackie, had been admitted to the Peter Mac Cancer Centre to have a lump removed from her breast.

I raced in and was told the surgery was a success and later that day my dad, John, and I visited Jackie in hospital. She looked pale and tired but otherwise okay. As a family we had just turned down the long and weary road of cancer treatment. More than 20 years on, Jackie has recovered and the cancer has not returned.

The treatment was rough and hard. So were people's reactions. When I told my friendship group – we were all about ten, old enough to know what cancer could mean – one of them screamed, she was so horrified. At the time I was angry at her reaction – *What a drama queen!* – but I am more understanding as an adult, such is the fear we all have of cancer. Our fear is well founded. People diagnosed with cancer in Australia between 2014 and 2018 had a 70 per cent chance of surviving after five years, meaning 30 per cent of those patients did not make it.[1]

Briefly explained, cancer is the cells of your body going off script. They multiply uncontrollably and, left untreated, will wreak havoc on healthy tissue. Incomprehensibly, the body is attacking itself. The question it left me with is – why? Jackie was a very healthy 48-year-old when she had her diagnosis; she had quit smoking many years before, rode a bike, ate more vegetables than anyone I knew and had a vibrant social life. Why would someone who was doing everything right get cancer? And why does cancer strike so many of us?

Early in my career as a science journalist, I was scrolling through journal articles hoping to find an exciting research story to break. A paper on evolution, cancer and Tasmanian devils jumped out at me as if it was in flashing lights.[2] Despite loving evolution, I had never thought about cancer through that lens. One of the co-authors, Rodrigo Hamede, was a researcher at the University of Tasmania's School of Natural Sciences, where I had completed a major in zoology just a few years before. He had never taught me, but I knew his reputation.

Intrigued, I got in touch with Hamede and mere minutes into the conversation I could feel my ideas about both cancer and evolution being picked up, turned around and completely reconfigured. So, revisiting the subject now, I went straight to Hamede to get the latest scoop on the origin of cancer and how evolutionary biology is paving the way for cancer prevention and treatment in the future.

The devil you know

Hamede is unusual in the world of cancer experts. An ecologist and epidemiologist with a career-long focus on the Tasmanian devil (*Sarcophilus harrisii*), you are more likely to find him in a pair of hiking boots and a tough rain jacket than a white lab coat. He has a long black ponytail, a short dark beard streaked with white, and an intensity in his eyes and bearing. In some ways he reminds me of his study subjects, and not just because he jokingly refers to his office as his 'den'.

The Tasmanian devil is the world's largest surviving carnivorous marsupial – that group of furry animals that raise their milk-fed young in pouches. They are about the size of a small French bulldog and almost as chunky. Devils roamed mainland Australia and the island state of lutruwita/Tasmania until around 3000 years ago when they disappeared on the mainland – most likely due to climate change, being hunted by dingos and pressure from humans.[3] Chilly, dingo-free Tasmania became the devils' stronghold, where they remained common – until recently.

Devils are lovable creatures in a rough and scrappy way. They are mostly black with a white stripe emblazoned on their chests, and pink pointed ears that show up a shocking red in torchlight. Young devils are cute, but adults quickly become gnarled and scarred. They gain their grisly aesthetic biting each other on the face in tussles for food and during courtship.

They are cheeky things. I distinctly remember returning to camp on a university zoology trip to find another student standing helplessly beside a ripped tent. We watched on as the plump backside of a local devil lolloped away with a salami. My classmate was beside himself; he was on exchange from

Europe, had borrowed the tent from our sternest professor, and he was down a sausage!

It is a good thing devils are plucky, as they are facing a great challenge. In the mid-1990s a female devil developed a cancerous tumour, most likely on her face or mouth. While fighting over a carcass or scrapping with a mate, she bit another devil. Then something strange happened. As she bit down, her cancer cells were transferred to her victim's wound. The unruly cells survived and grew on their new host as a parasite. Soon after this incident the female would have died, whether from the cancer or by other means. But her cancer cells lived on, spreading from devil to devil and killing individuals within months of infection. Often the infected devils died of starvation because the huge tumours on their faces prevented them from feeding.

The first hint of the epidemic came in 1996 when a photographer snapped a shot of an infected devil in the island's north-east. By 1999, devil scientist Menna Jones was catching infected devils further down the coast on the Freycinet Peninsula. Scientists scrambled to work out what was causing these disfiguring tumours, now named Devil Facial Tumour Disease or DFTD.

It was not until 2006 that researchers announced that DFTD was in fact a contagious cancer. The chromosomes in the cancer cells were 'grossly abnormal', but across individuals they were almost identical, showing that they came from a single origin.[4] Contagious cancers are rare, so this was explosive news. The only other vertebrate we know to be plagued by contagious cancer is the domestic dog. Dogs are vulnerable to a sexually transmitted cancer called Canine Transmissible Venereal Tumour, CTVT, which results in genital tumours. Shellfish also get transmissible cancers; some of these underwater cancers can even hop across the species barrier between bivalves.[5]

When I visit Hamede's Hobart office, he makes me lemon verbena tea with dried leaves from his garden and boiling water from a thermos that I am sure has got him through many cold hours of fieldwork. We sit and talk at a small table by his standing desk. On the wall is a map of Tasmania. Hamede has drawn the letter 'D' on all the places where animals with DFTD have been found. The map reminds me of a prop from a crime series where a killer is on the loose and detectives are trying to find a pattern to their crimes. Since the first known case in the state's east, DFTD has spread across the island. The Ds on the map reflect that spread, thickly scattered over 90 per cent of the animal's range with only the remote west yet to be impacted. While a lack of baseline data makes it hard to estimate precisely, in affected areas populations are thought to have crashed by 82 per cent.[6]

In 2014, wildlife vet Ruth Pye noticed that a tumour from a devil in the south of the state looked a little off. After the tumour was sampled for genetic testing, researchers were shocked to find that the cells were from a different pedigree. The Y chromosome in these cells showed the tumour had originated in a male devil, unlike the first – and most prevalent – cancer that came from a female. The fact that Tasmanian devils are afflicted with two of the three transmissible cancers that occur in vertebrates worldwide led to them being dubbed 'the world's most unlucky animal'. To Hamede, it suggested that devils may have encountered and survived transmissible cancers before.

Hamede started working with devils in 2004, early in the epidemic, and he was distressed by what he saw. 'I spent so much time in the field catching the same animals. You would see them in the pouch, then you would see them breeding, so you were attached to the individual. Then suddenly seeing them sick, deteriorating and dying was quite emotionally draining,' he explains. At this time there was a real fear that DFTD might completely wipe devils off the planet within decades.

Just as Hamede was losing hope for the species, he and his colleagues came across something miraculous: devils that were recovering from DFTD. 'It was 100 per cent mortality until we were lucky enough to see the first animal with a small tumour regression,' he says. This was in 2009, only 13 years since the first sick devil had been sighted. Was the species evolving to outwit the cancer? Soon, animals whose tumours were regressing were popping up all over the place. Hamede says: 'Sometimes the tumours shrink, but don't disappear, sometimes a tumour stops growing and the animal lives for another year and a half with the tumour not growing, and in some cases the tumour simply is just not there anymore.'

Tumour regression in humans does happen, but is vanishingly rare, with estimates from the 1960s suggesting it occurs maybe once in every 100 000 cases.[7] However, because of the phenomenon's rarity, these numbers are slippery. Spontaneous tumour regression is hard to study because of its infrequency. Additionally, there are ethical constraints in studying it in humans, as when cancer is detected there is a duty to treat it, so on exceptional occasions we may treat tumours that would have vanished anyway.

Hamede's devils with their disappearing tumours could provide insights into how species evolve to suppress cancer once it has taken root in the body. Testing shows that individuals with tumour regression have a specific immune response not shown by devils whose tumours do not regress.[8] He emphasises that these animals have 'managed to do something amazing'.

After some thought, he adds, 'I'm seeing the potential for the species to recover; the indications are good.'

But what the devil is cancer?

One way of understanding cancer is that it starts with one cell that does not want to play by the rules, often due to a spontaneous mutation of its genome. That cell multiplies and so do its offspring. Each new cancer cell contains genetic instructions to continue the uncontrolled growth. Cancer becomes clinically scary when cells become mobile, moving through the body to settle in new tissue – this is what is meant by metastasis.

Because cancer usually begins with a single cell that replicates out of control, it is tempting to think about it as a completely clonal entity. Not so. Many cancers are genetically unstable. Each cell division represents a generation, mutations occur regularly, and over the rapid-fire generations this leads to a diverse population of cells. These cells have different strengths, weaknesses, abilities and roles. In other words, cancer evolves.

A cancer's evolution is often directed when the body's defences take out some cancer cells but not others. The survivors are then harder for the body to deal with. This is natural selection, but it is working in favour of the cancer cells rather than a population. Natural selection is not the only force at play though. A cancer cell line is often unstable and mutating fast. Some evolutionary change within a cancer is just fuelled by random chance and not directed by selection, adding to its unpredictability.[9]

The devils give researchers a unique window into cancer evolution. As Hamede says, 'In a normal circumstance if an animal gets cancer, the animal dies, and the cancer dies.' But in the case of DFTD, 'Cancers can jump from host to host.' In this way the cancers gain a kind of immortality. In fact, the world's oldest cell line is CTVT, the transmissible cancer spread between dogs. The cancer can be traced back to a single individual that lived 11 000 years ago, and it is now spread across continents.[10] Because DFTD cell lines do not die with the host, researchers study it to get a glimpse of how cancer evolves over a long timescale, and how it responds to different threats.

Dinosaurs got cancer too

Discussion of modern carcinogens – things that cause cancer such as cigarette smoke, nuclear radiation, alcohol or pollution – can give the impression that cancer is a modern phenomenon. While all these things can significantly up the odds of an individual developing cancer, eliminating them would not eliminate the disease. In fact, cancer has ancient origins, and little hints of it can be found spread through the fossil record.

According to a 2016 study, 3D scans of a 1.7 million-year-old hominin footbone showed evidence of malignant osteosarcoma – a type of bone cancer.[11] The next most ancient example of cancer in the human family tree was from a 120 000-year-old Neanderthal fossil found in Croatia. The foot-bone analysis helped shift the way cancer is understood: rather than being a mostly modern concern, it is a disease that has plagued humanity from the beginning. Two of the researchers behind the paper, Patrick Randolph-Quinney and Edward John Odes, write:

> Cancer is often viewed as a fundamentally modern and monolithic disease. Many people think its rise and spread has been driven almost exclusively by the developed

> world's toxins and poisons; by our bad eating habits, lifestyles, and the very air we breathe.
>
> Actually, cancer is not a single disease. It is also far from modern. New fossil evidence suggests that its origins lie deep in prehistory.[12]

Cancer's malignant mark on the fossil record reaches far beyond early hominins living on an ancient savannah. A shell-less precursor to turtles, *Pappochelys rosinae*, died with bone cancer in its thigh bone 240 million years ago.[13] Dinosaurs got cancer too. The first dinosaur to receive the diagnosis was *Centrosaurus apertus*. This species looked like its famous relation the triceratops, and received the bad news in 2020, only around 77 million years after it had perished. The fossilised bones of *Centrosaurus* show evidence of significant osteosarcoma. Researchers found that the signatures of disease on the fossilised bones were strikingly similar to the bones of a 19-year-old human patient with the same diagnosis. In their paper describing this Cretaceous cancer, the authors noted:

> The extensive invasion of the cancer throughout the bone suggests that it persisted for a substantial period of the animal's life and might have invaded other body systems. A similarly advanced osteosarcoma in a human patient, left untreated, would certainly be fatal.[14]

So this crested, beaked dino died with – and probably from – an aggressive cancer. But was this a freak occurrence? Researchers have been eagerly examining dinosaur bones since the 17th century. As the first dinosaur cancer was definitively identified only in 2020, it would be tempting to think that cancer was rare in these ancient creatures. But the fossil record is fickle. To even form a fossil, conditions must be just so. And for cancer to be caught, conditions must be even more particular.

Cancers often occur in soft tissues that do not fossilise well, so it helps if the cancer is active in the bone. Additionally, specimens need to be well preserved. The animal also needs to have survived long enough with the disease for it to become evident in those bones. In a world where thundering predators like T-rex were just waiting to pick off sickly individuals, that is a tough ask.

Even so, there are many examples of neoplasms, tissue growths that can be benign or cancerous, in dinosaurs. But conclusively proving malignancy, i.e. spreading, has not been possible in these cases. Many of the tests that would be needed to conclusively determine whether a neoplasm was malignant are destructive to specimens, so to preserve rare fossils the tests have not been done. Modern dinosaurs – birds – do get cancer, which is another piece of data that points to cancer plaguing their extinct relatives. So just because ancient fossil evidence of cancer is scarce, does not mean that cancer was rare in the ancient world. The disease may stretch further into deep time than the world of shell-less turtles and sickly dinosaurs.

In the beginning, there was cancer

In the last decade some researchers have been pushing back cancer's first appearance to the dawn of multicellularity. We've met Athena Aktipis of Arizona State University and her research into the evolution of cooperation in human systems. She extended this passion into research on the multicellular body, and enthuses, '[The human body is] 30 trillion cells that are cooperating and coordinating every millisecond to make it viable for us to do all of the things that we do.' Her focus on cooperation theory has made her one of the key thinkers extending cancer's timeline by at least half a billion years, back to the ancient ocean.

Aktipis views cancer as a form of cheating within a cooperative cellular system. An urgent question for her was whether cancer existed across multicellular life. While cancer had been described across a range of species, no one had yet asked this question.

Taking the view that cancer was cells cheating within multicellular systems, Aktipis and her collaborators scoured the scientific literature for examples of cancer and cancer-like phenomena – cellular cheating without metastases – across multicellular life. Everywhere they looked, they found species where cancer had been reported. It popped up in mushrooms, red algae, brown algae, land plants and a vast array of animals. Cancer-like cellular cheating was even apparent in unicellular bacteria when they were living in cooperative colonies.[15]

Cancer's modern ubiquity implies an ancient past. Aktipis argues that it is very likely that the ancestors of each multicellular lineage all got cancer. As complex multicellularity has appeared at least six times in Earth's history, this also means that cancer has independently appeared in each of these lineages. The original 2015 research showed there were groups such as sponges and simple marine worms where cancer had not been reported. But since then, the search has been on, and Aktipis tells me this has revealed that 'cancer and cancer-like phenomena are pretty much everywhere and sometimes they hadn't been observed'.[16]

The idea that cancer occurs across multicellular life is not fully accepted within the field. Aurora Mihaela Nedelcu of the University of New Brunswick sometimes collaborates with Aktipis and colleagues, but when I ask for her opinion on their work mapping cancer across multicellular life, she tells me, 'I disagree.' Her concern? 'They are mapping the presence of tumours, and every cancer has to first have uncontrolled proliferation, but cancer [also] requires the ability of cells to migrate [causing metastasis].'

In life forms such as plants and algae, cells can proliferate out of control and cheat by not giving back to the multicellular organism. But those cells do not then move to other parts of the plant and continue their exploitation. Nedelcu says, 'If you go to the doctor and they find a tumour that has not spread, they're going to call it a benign tumour, they're not going to tell you that you have cancer.'

Nonetheless, Nedelcu agrees that understanding the interplay between multicellularity and cheating cells is vital to tackling the disease: 'The main question about the evolution of multicellularity is how to ensure that the cooperation is maintained in the face of defection, cheaters or selfish elements.'

Cancer's ancient history and connection to multicellularity deeply influenced Hamede's thinking. 'That has a huge implication,' he says, as we sit over our second cup of home-grown lemon verbena tea in his office. 'Multicellular organisms evolved *with* cancer.' This means that cancer has been with us since life started its journey to complexity. If we are going to game out how to respond to cancer, Hamede says, we need to understand what strategies multicellular bodies have evolved to coexist with it.

Natural selection is pretty good at weeding out traits that harm individuals; surely it should have fully cancer-proofed the bodies of modern organisms. This is one of cancer's many paradoxes. Given it is so deadly, why have we not evolved out of it? Aktipis's answer is that 'having some susceptibility to cancer is the price we pay for being multicellular organisms that can grow, that can adapt, that can respond to our environment, that can heal, that can be resilient and can reproduce'. Meaning that the only way to expel cancer from the planet would be to do away with multicellularity. She explains, 'There is an implicit trade-off in multicellular life where some capacities necessary for survival can leave the door open to cancer.'

Take wound healing, for example. To heal a wound the body needs to be able to quickly jack up the number of cells it is growing. But uncontrolled cell growth is a hallmark of dangerous cancer, so the body needs to be able to tell the difference between dangerous cell growth and wound healing in order to prevent one and promote the other. As Aktipis says, 'It takes time to evolve these sorts of complex information processing systems that can help regulate the body without leaving it vulnerable.'

Reproduction can also be linked to cancer. This can play out in a few ways. In some cases, the cell growth and hormones needed to reproduce can actively create an opportunity for a cancer cell to appear and take off, as in wound healing. Emerging research suggests that some genetic mutations that increase the risk of cancer also increase the number of children people have. In humans, mutations of the BRCA gene can increase a person's risk of breast, ovarian and other cancers. Studies comparing the number of children born to carriers of BRCA mutations and the number born to those without the mutations show that all genders with the mutation have a small, but significant, increase in fertility. This may have caused the gene to spread.[17]

Many cancers appear after reproductive age. This is because susceptibility to cancer can be passed down the generations. While sometimes cancer strikes early in life, there is a huge uptick in diagnosis after the average age of reproduction. This suggests that natural selection may not be as protective against mutations that increase the chance of the disease later in life compared to early on. If a parent carries these cancer-late-in-life genes or alleles, they can still raise kids before cancer strikes. This also means that an increased risk of late-in-life cancer is passed on to their children. Historically, when cancer susceptibility genes lead to cancer early in life, they do not get passed on, as the individuals with those genes do not reproduce. In this way natural selection is limited in its capacity to guard against mutations that lead to cancer after reproductive age.

But, despite having its limitations, natural selection has had half a billion years in the presence of cancer. In this time it has backed a diverse assortment of cancer suppression tools. Some of them might inspire the next generation of cancer treatment.

Wisdom from elephants

All this pondering of multicellularity, the history of cancer, conflict and cooperation can seem a bit detached from the hard realities of the disease. But there are researchers trying to develop new cancer treatments using exactly these ideas to inform their work.

The larger an animal is, the more cells it has. Logically, more cells should equal a higher chance one of those cells will go rogue. Within species this appears to be true. Taller humans have more cells than shorter humans and they are also more susceptible to cancer.[18] But does this rule stand up across species? No. If it did, whales and elephants would be dropping dead from cancer all over the place and mice and hummingbirds would be stunningly cancer resistant, but this is not the case.

The capacity for large, long-lived animals to suppress cancer despite their wealth of cells is called Peto's paradox, named after the statistician who first observed it, Sir Richard Peto. The paradox's capacity to inform medical research was laid out in an influential 2011 review by Aleah Caulin and Carlo Maley.[19] The authors stated:

> Animals with 1000 times more cells than humans do not exhibit an increased cancer risk, suggesting that natural mechanisms can suppress cancer 1000 times more effectively than is done in human cells.

Cancer therapies are often hard on the body, with lots of side effects. They take a long time, meaning patients like my mother who recover often spend years unwell. Despite decades of research, many advanced cancers remain incurable. These factors and more drove Caulin and Maley to argue that more research should be done on preventing cancer in the first place. And where should we turn for inspiration when it comes to cancer suppression? Species that are already really good at it, of course.

> Cancer treatments have not proven as effective as promised. If we can harness the cancer suppression mechanisms of large, long-lived organisms, then we could potentially eradicate cancer as a public health threat in humans ... People have been invested in cancer research for decades while evolution has been tuning cancer suppression mechanisms for over a billion years. It's time to learn from the expert.

Wow. This paper is coming in hot and making some big claims. While I think any future where 'we could potentially eradicate cancer as a public health threat in humans' is a very, very distant one, I do think their ideas are compelling.

The paper's lead author, Aleah Caulin, was a PhD student at the time it was published. After reading her confident, big-thinking words I was keen to see where she had taken her research 13 years down the track. Did her optimism hold? Chatting to Caulin over Zoom, I read her those quotes.

'I would still say that today,' she chuckles. The logic and optimism in that paper has had a cascading effect. Caulin is now part of a team developing drugs to try and supress cancer by using species that have better cancer suppression tools than we do as inspiration.

To see how she got there, we need to go back to 2008 when Caulin was in her first year of graduate school at the University of California, San Francisco. In a class run by Carlo Maley, a renowned cancer researcher now based at Arizona State University, she read a paper that discussed Peto's paradox, and it caught her imagination.

'I went back to my apartment and [thought], what's the biggest animal that's been sequenced?' she tells me.

There are databases online that contain the genetic codes of different species, which researchers can search for specific genes within those codes. It is like Google for genetics nerds with, admittedly, more steps than I am getting into here, but it is relatively simple to use. Luckily for Caulin, somebody had uploaded a draft assembly of the African elephant's genome. If we are talking land animals, that is about as big as it gets.

Animal bodies have multiple genes that code for proteins that help prevent and constrain cancer. A stalwart member of the human body's cancer defence team is a gene called TP53, which codes for a protein called P53. This powerful protein has the capacity to trigger cancerous cells to self-destruct, to stop cells dividing, and repair DNA. This action halts cancer in its tracks. TP53's fundamental role is underlined by the fact that many cancer cells have a mutation that inhibits P53 from forming.[20] The protein's role in warding off cancer is so important it has been dubbed 'the guardian of the genome'.[21] A prestigious title for a mighty protein.

After finding the elephant's genome, Caulin's next step was to search for known cancer suppressor genes. And found 'a ton of hits with P53', she tells me.

Humans have one copy of TP53. As we have paired chromosomes, this means we usually have two versions of the gene in each cell. Elephants have 20 copies, meaning they can have a whopping 40 versions of the gene. This initial finding was exciting: 'It seemed too obvious,' Caulin says, 'especially with P53 because it's such a crucial, crucial gene.'

Searching the database was a great start to Caulin's investigations, but to really understand what was happening in these behemoths' DNA, she needed their blood. Luckily the keepers at California's Oakland Zoo were happy to hook her up with the good stuff. After a complicated process of genetic wizardry, a solid picture started to emerge. One of the copies of TP53 that elephants have is very similar to ours and is thought of as the 'ancestral' TP53. This gene has heft; it is long and complicated. But the other elephant genes are far shorter. Small genes like this are called truncated genes – it's easy to remember: elephants have trunk-ated TP53. Are these smaller genes any good, or was something else going on? Enter Lisa Abegglen and Joshua Schiffman.

In 2011, the year Caulin's audacious paper was published, Lisa Abegglen joined Joshua Schiffman's lab at the University of Utah. Schiffman was diagnosed with Hodgkin lymphoma at 15 and since then has been on a lifelong mission against cancer. When he met Abegglen his driving questions were: 'Who's getting more cancer? And why?'

Abegglen joined the lab and started work on a rare condition called Li-Fraumeni syndrome. The ongoing role the TP53 gene plays in humans becomes devastatingly evident in this syndrome, which occurs when one of the two versions of the gene is mutated, impacting production of the P53 protein. This means P53 can no longer properly patrol the cell for DNA damage and stop cancer before it starts. This results in an incredibly high lifetime risk of cancer – greater than 90 per cent for women and greater than 70 per cent for men – with many cancers emerging in childhood.[22] Li-Fraumeni syndrome is heritable, and when one parent has it they have a 50 per cent chance of passing it on to their children. Although not everyone with the condition has received it from a parent: in 25 per cent of cases, it has arisen as a new mutation.[23]

Sometimes scientists have flashes of insight when suddenly an idea falls into place. This can happen when an apple falls from a tree, while sitting in a bath, voyaging through the Galápagos or in the collective consciousness of a lab. The eureka moment that changed Abegglen and Schiffman's lives came in 2012 while listening to Carlo Maley, Caulin's professor and co-author of that daring paper on Peto's paradox, at a conference.

Abegglen, Schiffman and their colleagues had been working hard trying to understand the difference in the capacity for DNA repair between people with two functioning TP53 genes and those with Li-Fraumeni syndrome. In the lab they found that having Li-Fraumeni syndrome compromises DNA repair – and fixing wonky DNA is key to keeping the body cancer-free. Abegglen tells me that 'Carlo Maley was speaking about

Peto's paradox, and he said that his graduate student Aleah Caulin had discovered that elephants have many extra copies of TP53. Here we are studying patients with one less copy [of TP53], and they don't respond well to DNA damage.'

Cue big idea. Schiffman still remembers the moment like it was yesterday: 'I almost fell out of my seat and a light bulb went off. That was the epiphany. In that moment, sitting in that auditorium in the dark, I thought, Oh my God, instead of looking at who is getting more cancer, what if we try to understand who's getting less cancer?'

After the presentation, Schiffman and Maley got talking. Schiffman and Abegglen had been deep in P53 research. What would happen if they used the tools they had been developing to understand how human P53 enacted DNA repair, to look at the elephant protein? Was elephant P53 operating in the same way, or in an even more effective way than our version of the protein?

A decade on and the results are coming in. One key finding is that the large version of elephant P53 is more effective than the human version at guarding against cancer.[24] This finding has recently been backed up by a separate team.

What about those truncated genes? Are they doing anything? Abegglen and colleagues added one of the truncated genes to human cancer cells in the lab. This gene was called TP53-RETROGENE 9, or TP53-R9 for short. They found it was tiny but mighty. Like its larger counterpart, it coded for a protein that made cancer cells self-destruct. Their findings suggest that this protein either teams up with ancestral P53 to increase a cell's capacity to take itself out in the face of DNA damage, or it has an additive effect.[25]

Schiffman and Abegglen were excited by the amazing capabilities of elephant cancer-suppressor proteins. But in the early days of their work, bridging the gap to human therapies seemed like science fiction to them. 'We can kill every single cancer cell in a dish. But getting that to work in a person or even in a mouse is more challenging,' Abegglen tells me. Luckily, within a few years they would meet Avi Schroeder of the Israeli Institute of Technology.

Schroeder makes genetic medicines delivered by tiny nanoparticles about one-thousandth the width of a human hair. These tiny vehicles deliver remedies to specific tissues in the body. He explains, 'I always think of it as a spaceship that carries an astronaut – the astronaut [being] the genetic material – into cells and then releases it inside the cells.' This was the kind of sci-fi mindset Schiffman and Abegglen needed.

When I first heard of this technology, I thought it must be a distant pipe dream. But similar tech is already in action to tackle other diseases. The best-known example of nanoparticles being used to deliver a genetic drug is the Moderna mRNA Covid vaccine. Like a waiter in a restaurant talking to a customer, mRNA gets instructions for a gene from the DNA, then travels with those instructions to where a protein is made, like the waiter going to the chef in the kitchen. The exterior of a nanoparticle is made of a lipid layer, similar to the membrane or outer layer of our cells. These liposomes contain instructions in the form of mRNA that tell the cells to start making the proteins on the outside of the coronavirus. In 2023 it was announced that similar technology has showed promising results in a trial of a vaccine-like medicine for pancreatic cancer.[26]

Schroeder and Schiffman also met at an academic conference. It was 2015 and it was scientific love at first PowerPoint presentation. Schiffman had the cancer-fighting proteins and Schroeder had the nanoparticles to deliver them. Schiffman tells me, 'It was like we were separated at birth, same sense of humour, same sense of urgency, same love of science.' Both men were in academia, a famously slow place to dream up new drugs, so they decided to merge their interests and start a biotech, which they named Peel Therapeutics, after the Hebrew word for elephant.

Their shared long-term dream is to use nanotechnology to create vaccine-like genetic medicines to aid in the prevention of cancer. Schroeder says, 'We could have a treatment similar to our flu vaccine – you get a once-a-year immunisation and the chances of getting cancer are much lower.'

So how exactly would their elephant-inspired therapies work? Schroeder tells me,

> The idea was that we take genetic material [TP53], either DNA or RNA, that the elephants have but we do not have … Then it could be loaded into a nanoparticle and then have it administered or targeted directly to the cancer sites. Then this material would actually be encoded inside the cancer cells, it would recognise the cell as cancerous and destroy the cancer cell.

Sounds simple, right?

A long list of challenges remain to be overcome before the dream can be realised. Researchers like Abegglen are steadily learning more about the truncated genes, but it is still uncertain which gene or combination of genes would be most effective. 'We'd probably try to look for the elephant gene that's strongest, most robust,' Schroeder says. Once the perfect gene is selected, the next step is to calibrate the right dose. Not enough and cancer cells may survive, too much and healthy cells may be destroyed.

Many modern treatments such as chemotherapies and radiation are indiscriminate, targeting both healthy and cancerous cells. This wholesale approach leads to nasty side effects and long-term impacts on the body. Schiffman, with his teenage history of cancer treatment, has experienced this personally: 'Two years ago, I had to undergo a triple bypass open-heart surgery at less than the age of 50 because of the radiation exposure to my chest and coronary artery.' Therefore, their focus is not just on making more effective drugs, but also safer ones.

Some new generation therapies are starting to be more specific. But the promise of elephant TP53 drugs lies in their double specificity. The nanoparticles themselves can search out cancer cells, and the medicine they deliver has been tuned by millions of years of evolution to target those cells.

The first drugs Schroeder hopes to create using elephant tumour suppressors will not be universal vaccines. He envisions the first treatments being for people who are at a high risk of cancer, such as those with Li-Fraumeni syndrome or BRCA gene mutations. These therapies could be used to reduce cancer risk after serious exposure to carcinogens or after a cancer has gone into remission.

For Abegglen and Aleah Caulin, elephant TP53 was just the start of their journey into medicine inspired by the mechanisms other species have evolved to avoid getting sick. Abegglen still works with Schiffman, and as well as her ongoing work with TP53 she is now looking at how other surprisingly cancer-resistant species fight off the big C.

Researchers analysed 6049 necropsy records for 292 species from 99 zoos for a massive collaborative paper Abegglen worked on that was published in 2023. They found that the common porpoise, Rodrigues fruit bat and the black-footed penguin were all incredibly cancer resistant, with neoplasms found in less than 2 per cent of their bodies, while ferrets seemed incredibly cancer prone, with 63 per cent of their bodies harbouring neoplasms.[27] For Abegglen this lights the way for exciting new avenues of study. 'I absolutely believe that there are secrets to fighting cancer, and maybe even preventing cancer, in nature,' she says.

Caulin is now working at the Peel Institute with Schiffman and Schroeder. She has witnessed a massive shift in the way cancer researchers think about evolution. Fifteen years ago, when she was watching Schiffman give talks on the topic to cancer biologists, he would have to really go back to basics with the audience – evolutionary theory was not front of mind at all. Today, when Caulin explains these ideas to colleagues, it is not seen as wacky. 'It's still really cool,' she says, 'but less crazy.'

Using evolution to beat cancer at its own game

New whiz-bang therapies are an exciting prospect, but what can be done now? When cancer outfoxes a species' cancer-suppression mechanisms, it starts evolving in the body. Robert Gatenby and his colleagues at the Moffit Centre in Tampa, Florida, are using evolutionary theory to try and use today's technologies to squash advanced cancers. The team takes direct inspiration from how evolution drives change in animals in nature, then use those lessons to innovate better treatment regimens.

Most cancers can be cured if they are caught early enough. But left untreated, cancer cells multiply and spread. Some metastatic cancers are survived by a fraction of patients, but usually when a cancer has metastasised it is incurable. Survival rates depend on many factors such as the type of cancer and the treatment available. Tackling incurable metastatic cancers is Gatenby's main aim, with some of his most exciting work being done on incurable, metastatic prostate cancer.

In 2020 prostate cancer was the cause of death for 3568 Australians, ranking it as the third most likely cancer to kill in that year.[28] In its early stages it is curable, but once it metastasises dramatically beyond the area surrounding the prostate, it is universally deadly. An incredible amount of money has gone into developing new treatments for incurable prostate cancer. I can hear the frustration in Gatenby's voice when he says, 'The embarrassing thing is that the fraction of men with metastatic prostate cancer that are alive at five years is lower now than it was in 1977.'

This dismal history suggests to Gatenby that something is seriously wrong with a lot of research. He wants to fix this and dreams of turning metastatic cancer from a disease that is nearly always deadly to a chronic disease that can be managed like diabetes, or even eliminated.

He emphasises to me that understanding how cancer evolves in the body is crucial if we are to improve outcomes for cancer patients, arguing that 'we could cure cancer if it weren't for evolution'. This is a big claim. For decades drug companies have thrown billions of dollars into new drug trials, but Gatenby does not believe it is the effectiveness of the drugs holding treatment back: 'We have a lot of very good treatments for most cancers. [In the case of incurable metastatic cancer] these treatments often drive the cancer to the point where it's invisible, yet it always comes back. And so, evolution defeats us.'

Traditionally, metastatic cancers and patients alike are smashed with the maximum continuous dose a patient can tolerate until the cancer starts growing again. At this point an oncologist may switch up which drug or treatment is given. This strategy provides the perfect environment for cancer cells to evolve resistance to treatment.

Cancer cells by their nature are very diverse, both in their genetics and in the way they operate. This means it is likely that some will be able to survive any given treatment. If even a few cells, often impossible to detect, survive a treatment, they will eventually start to multiply. This is natural selection at work and means the whole new population of cells is now resistant to that treatment. When a new drug is applied, the cycle begins again. This predictable journey to treatment resistance told Gatenby that the traditional approach to metastatic cancer needed to be challenged.

Could cancer be managed like diabetes?

Some cancers are incurable with the technology and skills we have today. Despite this, treatments for patients with an incurable disease have 'an implicit goal of cure', Gatenby says, 'when you know quite clearly that that's not a possible outcome'.

The impact of this bullish approach on a patient's quality of life upsets and frustrates Gatenby. He and his team are trying to flip the script. When a cure is not possible, they are developing treatments to extend life and increase patient welfare inspired by how populations evolve in environments. A pilot trial of one of these treatments, dubbed adaptive therapy, has shown promising results for patients with metastatic prostate cancer – but it comes with a catch.

Adaptive therapy is not trying to eliminate cancer from the body. Rather, the goal is to control it and prevent its progression. The cells that survive treatment at the maximum tolerated dose are not only resistant to treatment but find themselves in a cancer-cell paradise. Before the drug, they had to compete with the non-resistant cancer cells for the body's resources. But now the slate has been wiped clean. These resistant cells have been freed from constraints and now can run amok in the body in a phenomenon known in ecology – the study of organisms in their physical environment – as competitive release. Adaptive therapies use lower doses of drugs to keep a controlled population of non-resistant cells around, but managed. This is a strategy familiar to many politicians – if your enemies are fighting, they are less of a threat to you. In this way Gatenby hopes to transition cancer from a deadly disease to a chronic one that can be managed with drugs, like diabetes.

Between 2015 and 2019, 33 patients with metastatic prostate cancer were enrolled into an adaptive therapy trial run by Gatenby and his colleagues Jingsong Zhang, Jessica Cunningham and Joel Brown. Sixteen patients were given 'standard of care' (SOC), the standard therapy, where they were given the maximum tolerated dose until progression. The remaining 17 trial participants chose to embark on a course of adaptive therapy.

When I spoke to Gatenby, the results were already stunning.[29] Sadly, all the SOC patients had passed away, but four patients of the 17 in the adaptive therapy group were still alive. Those who passed while having adaptive therapy lived for an average of 58.5 months, significantly longer than their SOC counterparts, who survived an average 31.3 months.

How did Gatenby and his collaborators pit the cancer cell populations against each other? Prostate cancer needs to be in the presence of testosterone. A common prostate cancer treatment is abiraterone, which prevents the body making the hormone. 'It's called chemical castration … men hate it,' Gatenby says. It is possible to measure tumour growth or decline by analysing a protein in the blood called prostate-specific antigen, or PSA.

In the trial, adaptive therapy patients would take abiraterone until their PSA levels dropped to less than 50 per cent of their original baseline. This was enough to reduce the tumour size but keep enough of the treatment-responsive cells around to keep those resistant cells in check. They would only start the drug again when the PSA returned to the baseline, indicating the tumour had grown. This cycle continued, never letting either population of cells go rogue and giving the patients a break from the medication and its side effects.

Treatment for each patient was mediated using a mathematical model, the Lotka-Volterra model, which is used widely in evolutionary ecology. Gatenby tells me, 'The mathematics is nowhere near novel. It's standard stuff.' Without getting into the detail of the mathematics, the thing to know is that the model helps the oncologists understand the population dynamics of the cancer cells and anticipate when to apply the drugs.

Using a mathematical model also allows for each regimen to improve with each patient. The data from patients can be fed back into the model and used to update it. This allows the drug trial to have a level of granularity that trials that simply measure one treatment against another do not.

'We could look to see what we did right, but we also learnt what we did wrong,' Gatenby tells me.

This use of mathematics allowed the researchers to identify a design flaw in their experiment that led to some patients being overtreated. The tumours were being monitored by PSA levels every month, but before treatment could stop, tumour shrinkage had to be confirmed using a CAT scan, which was only available every few months. The patients who were overtreated had PSA levels below 50 per cent of their pretreatment value well before their CAT scans were booked in, but they had to keep getting treated until their tumour regression was confirmed by the scan.

While these patients were still off treatment significantly longer than those under the SOC, they were on treatment for longer than those patients whose scans and dropping PSA levels synced up. While they were waiting for their scans, too many of the treatment-sensitive cells were killed off, allowing for competitive release of resistant cells. Patients who were lucky enough that their PSA levels matched up with the dates of the CAT scans received just the right amount of treatment. They are the patients who are still alive as I write this, massively out-surviving the standard prognosis for their condition.

Unpicking what happened here has changed the researchers' approach. Gatenby tells me that he has learned: 'We have to be very careful, we need to be monitoring the patients, and as soon as they get to 50 per cent, we've got to stop [treatment] and not even wait for them to see the doctor.' Integrating this new data into his models gives him huge hope for the future of adaptive therapy. He says, 'Our math models suggest we could control tumours for 20 years with this approach, and since [metastatic prostate cancer is] a disease of [primarily] older men, I think we could effectively eliminate it as a cause of death.'

A cancer extinction event

While controlling cancer is great, the ultimate hope is for a treatment that will eliminate it altogether. In other words, make cancer go extinct. Gatenby muses that 'In the popular imagination, we always go to the dinosaur when we think about extinction.' A comet or asteroid slammed into the Earth 66 million years ago, kicking off the Cretaceous–Paleogene mass extinction. This was a dramatic event in Earth's history. Chaos reigned for years after the initial impact, leading to the disappearance of at least three-quarters of the world's species, if not more. Gatenby likens the Cretaceous–Paleogene mass extinction to a cancer patient being treated with the 'maximum tolerated dose'.

The problem with treatments that mirror this incident is that they are indiscriminate. Dinosaurs were not the only creatures to fall in that extinction event. Less charismatic plankton took a massive hit, many lineages of marine molluscs did not make it through, and graceful pterosaurs never flew again.

These losses are akin to the damage sustained by normal cells during aggressive cancer treatment. No one wants to put their body under that kind of pressure to kill off cancer cells. Maximum tolerated doses and giant space rocks 'are really big evolutionary forces being applied', Gatenby says.

If the Cretaceous–Paleogene mass extinction was a conscious effort to destroy dinosaurs, we now know that it would have failed. A small, unassuming group of feathered dinosaurs survived: the birds. As birds are the class of land vertebrates that survived with the most species described, this failure is spectacular.

Clearly Gatenby needed a better model of extinction. Strangely, the extinction crises we are currently witnessing have provided him with a plethora of valuable examples. The fate of the non-avian dinosaurs is not the 'standard way' species go extinct, and Gatenby believes that modern extinctions can be replicated in the body to make metastatic cancer extinct.

One of his muses is the heath hen (*Tympanuchus cupido cupido*). This unusual-looking bird was declared extinct nearly one hundred years ago. A North American grouse, it was a bizarre, mostly land-loving bird with an unfortunate attribute that made it vulnerable when Europeans arrived on the continent: it was tasty.

By 1870 the once diverse and widely distributed subspecies was restricted to an estimated 50 birds living on Martha's Vineyard. Locals protected them and soon there were about 2000 heath hens on the east coast island. Things were looking up until disasters started to rain down on the small population. In a 2019 paper, Gatenby and colleagues detail the catastrophes that led to the hens' extinction: 'A fire destroyed part of their breeding area, several winters were unusually harsh, and an infectious poultry disease appeared.'[30] None of these events would have knocked out the species before its distribution and diversity had been so diminished, but now they spelled the end of the heath hen.

How does the hens' demise map onto possible treatment regimens? Horrifyingly, one gram of tumour can be made up of one billion cancer cells. By the time someone has incurable metastatic cancer they can have over 100 grams of tumour in their body. As the cancer cells spread and multiply, they mutate, creating diversity. In his 2019 paper Gatenby writes that the cancer cells in a patient with metastatic cancer are 'roughly equivalent to the size and diversity of the global mouse population'.[31]

While most species that have gone extinct in the last few hundred years have not had populations nearly as vast or diverse as the global mouse population, or the cancer cells in a very unwell person, they do have a lot in common with both. Many species that have recently been driven to extinction were once diverse, had large populations and inhabited vast, connected ecosystems. Their downfall began when their worlds fragmented and populations were significantly reduced by human activity. Their isolated populations then became vulnerable to random small events that previously would not have massively impacted them. Gatenby has been figuring out how to mirror these extinction events within the human body.

So far, his work gaming out how to engineer cancer extinctions in humans has been theoretical. But he is about to put his ideas to the test. As I spoke to him in August 2023, he was actively recruiting participants for a trial targeting incurable pancreatic cancer that would employ treatment methods inspired by anthropogenic extinctions.

These patients' cancers started in the pancreas and spread throughout the body creating large, diverse populations of cells. Gatenby's idea is to go in hard and fast with a treatment to massively reduce the population of tumour cells. In regular cancer treatment this is when oncologists would usually back off, let the patient recover and only resume treatment or switch things up when the cancer is detectable again. But, as Gatenby says, 'The problem is that once you can measure it … there's billions of cells, it's a massive population and so it's too late.'

Extinction treatment does not wait for the cancer to regroup, it goes straight in with a new therapy. Gatenby tells me a colleague describes traditional cancer treatment as 'like a boxing match … we knock it to the mat, and then we go to our corner, wait for it to get back up again'. But it needs to be looked at more like a knife fight 'when you've got your opponent on the mat. You close and kill it'. Such an onslaught of drugs and treatments sounds hard on the body, but because after the first massive dose the populations of cancer cells are small and less diverse, the subsequent therapies can be applied at a lower dose. Just like the heath hens on Martha's Vineyard, the reduced populations of cells are vulnerable to 'small perturbations' that they could withstand when the population was larger and more diverse.

Listening to Gatenby describe extinction protocols, I was reminded of the therapies my mother underwent in the early 2000s. She began with surgery, then had rounds of chemo, followed by radiotherapy and eventually a hormonal therapy. It was a long road, but the sequential application of drugs sounded very similar to what Gatenby was proposing. I describe Jackie's therapy to him and ask if he is reinventing the wheel.

No, not quite. My mother's cancer had not spread throughout her body. Breast cancer can be cured if it's localised, 'But as soon as that breast cancer becomes metastatic, it's fatal, that transition is really important,' Gatenby explains. His primary focus with extinction therapy is applying it to metastatic cancer, which luckily my mother did not have.

But localised cancer still comes with dangers. After having her tumour removed there was an assumption by the oncologists that some small, undetectable populations of cells remained in my mother's body. So the therapies she had after surgery aimed to eliminate them. This is common and called adjuvant therapy. And, as I witnessed, those additional long, sequential treatments are tough on the body. They are also not foolproof, with many people going through the struggle of treatment only to have their cancer bounce back, now drug resistant and possibly metastatic.

In some ways adjuvant therapy is a precursor to extinction therapy, as it uses some of the same logic.[32] Various treatments are applied with the hope that if cancer cells do not respond to one attack, they will be vulnerable to another. But Gatenby believes adjuvant therapy could be greatly improved if the frame of anthropogenic extinctions was used to optimise it.[33] Currently, he says, for many cancers oncologists give the 'treatment over and over again for six months, and if they [want to] intensify therapy, they simply give it for longer'.

This was true for my mother, who was on chemotherapy for months.

Gatenby and his colleagues want to shake up the accepted wisdom when it comes to adjuvant therapy. As with their therapies for metastatic cancer, they take inspiration from anthropogenic extinction. These therapies would involve giving treatments as early as possible after diagnosis and after another treatment has finished. So there is treatment cycling between different agents. Cycles should be short, and quickly followed by a new cycle. And the vulnerabilities of specific cell populations should be considered.

If my mother's treatment was like this instead of long months on chemo, she would have been flipping through smaller cycles of a range of treatments, each given to specifically target any potential cancer cell that may have hung on. In theory, Gatenby's ideas for improving adjuvant therapy would also have meant that she was on lower doses and the treatment would have wrapped up sooner.

But could this actually work? While small trials are underway with his protocols, Gatenby's work is very cerebral. Will the theory translate into practice? Well, it has for the treatment of childhood leukemia. The treatment regimen for kids diagnosed with pediatric acute lymphoblastic leukemia closely resembles his extinction regimen. Patients are given a treatment until the cancer is undetectable. But the assumption is made that some cancer cells will survive, so another treatment is quickly given, without giving the cancer any time to recover. This is incredibly effective, and many children diagnosed will survive their illness.[34] Bizarrely, this method of treatment was not inspired by Gatenby's evolutionary lens, but through another experimental pathway. But it is successful and follows his ideas to a tee.

Can our approach to cancer evolve?

Listening to Gatenby evangelise about adaptive therapy and his extinction trials, it is hard for me to keep up my journalistic detachment and not get carried away with excitement. But something is niggling at me. Especially when it comes to adaptive therapy, Gatenby has had some of these ideas for 20 years. Cancer is such a horrific global scourge, if his approaches were the answer, surely they would be the standard of care by now. So what is holding his research back? Is it flawed? Gatenby provides big reasons why his ideas have not gone mainstream sooner:

1. It is against the financial interests of the medical industry.
2. The idea is new and scary.

The drugs that Gatenby is proposing to use are not new, nor even on patent. Therefore, he believes, drug companies are not incentivised to support his work. His trials have had to be run on a shoestring budget. This trickles down to how willing cancer centres are to support the work. Gatenby tells me, 'We've had people that want to do the trial in other places and can't because it would compete with drug company trials, and to a cancer centre a drug company trial is worth way more.' Even if new on-patent medication was being used, the quantity of drug is reduced, making treatment cheaper. Gatenby estimates that the patients on the prostate cancer trial were around US$70 000 cheaper to treat than patients on SOC: 'To me, that's what was really good news, but to a cancer centre it is not good news because they're the ones losing money.'

Worried I might be falling into an anti-big-pharma conspiracy trap, I set about asking cancer researchers and medical professionals whether Gatenby's claims about why his work has been stymied stack up. They have all told me, with exasperation, that Gatenby's predicament sounds very real. Saskia Freytag, laboratory head at the Brain Cancer Research Lab of the Walter and Eliza Hall Institute of Medical Research, tells me, 'No pharma company is going to fund this.' Freytag believes government funding would also be 'hard to come by', especially as our system sets up pharmaceutical companies to be the primary funders of trials. She identifies philanthropy as another option, but also warns it would be 'pretty tough, as the risk of being associated with a clinical trial is high'.

There are still questions to be answered before many clinicians will comfortably employ adaptive therapy, although there are a few individuals who have successfully advocated for adaptive therapy with their oncologists. Freytag noted to me that there is a general trend towards lowering doses outside of the framework of adaptive therapy. She is interested in understanding whether the results were due to the use of the evolution-inspired model or just the result of a lower dose.

It will take a combined effort of government-funded bodies, non-government organisations, universities, research groups and private funders for adaptive therapy and extinction therapy to get the rigorous testing they need before clinicians can be confident in using them. In 2023 a trial using Gatenby's methods started enrolling patients.[35] This trial is being run across Australia, New Zealand and Europe and is a collaboration between three entities who received mixed funding from the above-mentioned sources.

How about Gatenby's other claim, that his ideas are too different for people to accept?

Joshua Schiffman of Peel Therapeutics believes that Gatenby's style of evolutionary medicine faces an 'uphill battle because it's a culture shift ... People are wary of new ideas, and this is counter to what we're all taught in medical school, which is you have to roll up your sleeves and stamp out disease'. He also believes that adaptive therapy can challenge pharmaceutical companies and researchers alike: 'We have billions of dollars going into clinical trials and everyone's built their career on trying to get rid of cancer.'

Gatenby was anxious as he embarked on the prostate cancer adaptive therapy trial. He had done the theory but constantly questioned himself: 'His idea of these two resources competing against each other, does that happen or not? ... I was terrified that we would actually do some harm to somebody because we didn't know for sure it would happen.' This nervousness was shared by the patients.

Finding patients who were willing to harbour cancer cells within their bodies was one of Gatenby's major challenges in embarking on the trial. More than 50 years of cancer research and treatment has centred on an all-out-war approach, so making peace with the idea of cancer in the body is not easy. If Gatenby's ideas are going to be successful, we will have to radically reimagine cancer. To do that, it might be helpful to look to other species, and return to the the Tasmanian devil discussed at the beginning of this chapter.

Hope at the end of the road?

Competition and cancer are front of mind as Rodrigo Hamede, a few of his PhD students and I bump along a dirt road in a four-wheel-drive on a rainy November day. It is freezing even though summer is around the corner. We are on the borderlands of DFTD1, the strain of contagious cancer that appeared in the 1990s, and DFTD2, the 2014 version. Originally DFTD2 was contained within a stubby peninsula in southern Tasmania, but it is now leaching out into new territory. Our mission? To check out devils who are living on the advancing frontline of DFTD2. The population has been living with the earlier cancer for quite some time and is being actively monitored by Hamede and colleagues. The new disease's spread provides the researchers with a unique opportunity. This is the only place in the world we know of where two types of transmissible cancers have met in the same host. Hamede and colleagues are getting a front-row seat to how the devils and DFTD1 and DFTD2 react.

Ground zero is a misty property nestled among green forested hills. It is in Sandfly, just a 20-minute drive from Hobart. As we roll up the driveway and stare at the mountains behind the property, we could be hours away from any town. Such is life in Tasmania. The couple who own this place have allowed Hamede and his team to trap devils as part of their ongoing monitoring. They join us as we traipse down through a paddock, over a small stream and past a grove of tree ferns to where the forest begins.

During the drive Hamede filled me in on why today's work is significant. He is hoping to find out whether the two types of cancer are co-existing or competing. He suspects 'there is competition, competition for a resource, and that resource is the Tasmanian devil … I expect that this competition is going to result in changes in certain traits of the tumours'.

How these cancers will evolve as they compete is yet to be seen. Hamede expects that the changes will relate to transmissibility, virulence, and the time between infection and the animal becoming symptomatic. All these factors make the cancer more likely to be passed on to another animal before the original host dies. Cancers that are highly transmissible, grow fast and make themselves known early, might be more likely to survive.

A PhD student calls out to let us know there is a devil in one of the tube traps. A heavy hessian sack is placed around the end of the tube and the devil is released into it. Calmly sitting on the wet ground in waterproofs, Hamede manoeuvres the animal and the sack so the devil is facing the opening. She's a female, just a year old. All I can see is her open mouth as I clutch my phone camera and think about the fact the devils' bite force is the strongest in relation to body size of any animal on Earth. Yikes. I am getting excited and nervous as Hamede starts his inspection. Will I get to see her close up? Do I even want to if she is sick?

'No tumours,' Hamede calls out. What a relief. Gently he takes a small biopsy of her ear and a blood sample. Hamede invites me closer as he adjusts her again to inspect the pouch. I lean over and stare into it as with gloved hands he pulls the elastic skin of the entrance apart. Inside it's pink and, bizarrely she has ... breasts. They are unmistakable – two round enlarged bits of tissue with nipples atop. Opening the pouch further he points out other tiny unused nipples. Does this mean she is caring for devil joeys, imps? 'She has two,' he says, warmth in his voice. The two breasts showed him that back in a den somewhere were young she was regularly feeding; if she had a larger litter, more of her nipples would be engorged.

Maturing and growing faster is one of the ways devils are evolving to survive DFTD. Females used to start breeding in their second year, but in populations impacted by the disease, one-year-old mums are common. Devil behaviour is also changing, with bites becoming less common, especially among older devils. However, when young devils do bite each other, their bites tend to leave deeper wounds compared to those of old souls, which is thought to increase the odds of contagion. This is lightning-fast evolution, but as the disease is so deadly, there is a massive selection force giving a huge advantage to individuals with any traits that are protective against the disease or help them breed before it takes them down.

It is time to release this young mum back to her responsibilities. We all stand well back as Hamede manoeuvres the animal in the hessian sack again. For the first time her whole head is popping out and we see her. She is perfect. Older devils can look scrappy and scarred, like members of a particularly rough motorcycle gang. But she is young, fresh, with the jet-black face, pink nose and two distinct white eyebrows. Hamede turns to the property owners. 'That's a lovely creature you have on your property,' he says before opening the bag and letting her escape. She bounds off into the forest, presumably making a beeline for her imps snuggled up in a den somewhere.

After a few hours of freezing in the field, we return to Hamede's office to warm up, have another cup of lemon verbena tea and debrief the day. On the drive back my mind was racing. What would we have done if she had been sick? Would we have had to euthanise her? What of her little imps? That depends, Hamede says. If we had caught an 'emaciated devil that was dying in our hands', we would have needed to euthanise it. But if our young mum had a tumour but could go about her life, she would have been released. 'She's got two young in the den, she's got to wean them, she has things to do ... The welfare issues for the future generations are as important as this one.'

The argument against euthanasia goes further than getting one mum back to the den. It is wild devils living with cancer that have been evolving mechanisms to coexist with, and in rare cases beat, the cancer. 'The last thing you want to do is to alter that evolutionary trajectory,' Hamede says. How much to intervene with the devils and their cancer is a contentious issue in the field. Despite the exciting changes being observed in the species, the disease continues to spread among the population, threatening the devils' survival.

Remember those devils that have gone into spontaneous tumour regression? This has now been observed in over 50 devils. The individuals with tumour regression whose samples were analysed showed a single point mutation of the RASL11A gene, a gene linked to tumour suppression in some human cancers.[36] 'We don't know the full story,' Hamede says. 'If we knew we would be working with a lot of oncologists around the world ... but we do know that there is a genetic basis for those animals with tumour regressions.'

Studying the African elephant genome, Caulin was able to identify how the species suppressed cancer, but exactly how that gene was selected for over evolutionary time is lost. What I find exciting about the devils is that we are getting to watch that selection event play out in real time. What is more, selection seems to be operating on a different gene, adding richness to our understanding of the body's cancer defences. As the devils continue to evolve, I am confident that their story will directly impact cancer treatments. But the species must survive if that is to happen.

Being carrion eaters, devils love to gorge on roadkill, meaning they are often squashed mid meal. They are also losing habitat to development and forestry. To save them, significant effort needs to be put into reducing these risks. DFTD has reduced their population and these other 'perturbations' could threaten them in the same way that Gatenby hopes to eliminate cancer with small diverse applications of treatment once a population of cancer cells is reduced. As Hamede says, 'The conservation future of the devil relies on their own evolutionary potential and our ability to foster that evolutionary potential and reduce all other threats for the species.'

CHAPTER 6

Why do we age?

As I am writing, Sadie the miniature schnauzer is snoozing behind me. She is just over five years old and has been my companion on this writing journey, jumping into my lap when I interview scientists over Zoom and coming to get me when it's time to take a break and have a walk. At five, she is very much an adult – cheeky, but all grown up. I first had the idea for this book in 2019. At that time she was just a puppy, full of beans, still learning the ways of the world and growing fast. She had not quite mastered the art of snoozing while I worked.

If Sadie were a five-year-old human, she would still be a child. But as different species, she and I are at similar stages in our ageing process. In the slightly unscientific accounting of dog years, she is in her thirties, like me. I hope to still be fit and active at 40. When I hit that milestone, Sadie will be slowing down, approaching true old age. If she lives to 15, we will say she 'lived a good long life', whereas I can hope to live into my 80s or beyond. Even more dramatic differences can be observed looking at other species. A mouse is considered ancient after a few years, but Fred, a 110-year-old sulphur-crested cockatoo who lives at an animal sanctuary near me, was still going strong last time I visited.

With some rare and intriguing exceptions, ageing is universal across species. As with cancer, it seems odd that natural selection would have allowed a process to persist that is so incompatible with continued reproducing – why has it not been selected out of existence? And why do some species age faster than others?

Cancer is a complicated beast but ageing takes complexity to another level. It is a multifaceted system with many drivers. Evolutionary biologists have not reached agreement on how ageing – or, as they like to call it, senescence – evolved. This creates an urgent yet exciting research landscape. It is urgent because age-related diseases cause immeasurable suffering, and understanding ageing might offer ways we can live better lives for longer. It is exciting that in this world of artificial intelligence, space travel and unimaginable technology there are still major scientific questions to be answered.

Why tackle ageing? Alongside everything we do not know, there are a lot of compelling ideas that go part of the way to answering why we age, and why we do so at different rates. Many of these hypotheses directly link environmental conditions to the evolution of animals' life spans. So as humans rapidly change the Earth, it is important to know how these changes affect the patterns of other species' lives as well as our own. Additionally, we are at a time of great medical and cultural change that could radically shift what ageing means in the not-too-distant future. As we prepare to make decisions about the future of ageing, it is helpful to look at what it is and how it evolved.

What does it mean to 'get older'?

Ageing is a little bit like pornography. We know it when we see it, but it is tricky to define. Is it wrinkles, grey hair, loss of mobility, gaining of wisdom, a sudden interest in bingo? Is it vulnerability to disease, reduced capacity to recover from illness or injury? In 2020 Marios Kyriazis suggested that ageing be thought of as time-related dysfunction.[1] Kyriazis is a medical doctor and evolutionary researcher at the UK-based ELPIs Foundation for Indefinite Lifespans. His definition has the dual benefits of distinguishing it from other things that could slow us down, like an injury or infectious disease, and allowing for personalisation, as everyone has different things they need to feel functional.

A gloomier definition of ageing is that it represents an increased risk of death over time, especially from adulthood onwards. In modern human populations this plays out as mortality doubling every six to seven years among adults. It is a blunt measure, but useful when trying to understand ageing in large populations, especially of wild animals where assessing levels of dysfunction might prove tricky.[2]

What's happening in the body? A more technical way to describe ageing was developed by Carlos Lopez-Otin of the University of Oviedo and colleagues and launched in a 2013 paper.[3] The paper broke ageing down into nine 'hallmarks', each representing an element of ageing. This organised, detailed, and nuanced portrait of the process took the ageing research community by storm. The hallmarks disentangled this complicated and biologically messy phenomenon into parts that could be independently examined. For many scientists, this was a dream come true.

These famous hallmarks are broken down into:

1. Causes of ageing. These hallmarks involve things going wrong genetically.
2. Responses. The body must respond to these genetic challenges.
3. Results. How 1 and 2 impact the body.

Let's meet them.

Causes of ageing: genes gone wild

Mutations of the DNA in the body's cells can reduce the capacity of those cells to function. An extreme example of mutations messing things up is cancer. DNA mutation was once thought to be a major driver of ageing. It is a driver of cancer, an age-related phenomenon, but random mutations are no longer enemy number one in ageing research.

At the end of each chromosome – the long strands of DNA in each of our cells that condense when cells replicate – are sections of DNA and proteins called telomeres. This DNA is not there to code for proteins. Instead, its job is to protect the rest of the DNA during cell division. With each cell division the telomeres take a hit, becoming shorter and eventually losing function, causing genetic chaos and ageing.

DNA provides the codes to make proteins, which then carry out functions in our bodies, however, sometimes things can go wrong with the proteins themselves. For example, they can be the wrong shape or clump together. This can be associated with diseases like diabetes and Parkinson's disease. This is called loss of proteostasis.

Most cells in our body have the same genetic code. It's just that a skin cell is using that code to be a skin cell and a liver cell is using it to be a liver cell. How do they know what to do? The answer is an organisational system that would make any librarian proud. The DNA is packed and labelled in specific ways so that only certain parts of the code are read at different times. If a naughty kid came and changed the labels on the shelves in a library, it would be hard to find the book you need. As we age, the organisational system, called the epigenome, can get mixed about. This can result in the wrong bits of DNA being read out. Epigenetic alterations increase over time, reducing the capacity of the cells to do their jobs.

Responses

These next hallmarks are thought to occur as responses to the 'causes' of ageing and might be the result of functions that were helpful in the beginning but over time become a problem.

Living things need to consume specific amounts of certain nutrients to survive. The capacity to sense these nutrients gets dysregulated with age. Nutrient-sensing dysregulation reduces a cell's capacity to respond to its surroundings.

Mitochondria, the powerhouses of the cell, liberates energy for the rest of the cell to use. These powerpacks get dodgy over time – which is a problem, as they provide cells with energy to function. Mitochondrial dysfunction can also result in the structures leaking toxic substances into the rest of the cell.

Some cells in the body will stop dividing into two new cells as they should. But instead of dying gracefully, they lurk in the body, becoming 'zombie cells', not functioning and slowly leaking harmful chemicals. This is called cellular senescence.

Results

Our body is a tight collaboration of communicating cells. Cells of specific types work together to form tissues, which in turn make organs, and so on. All this chaos results in altered intercellular communication. After things go wrong genetically, nutrient-sensing and mitochondrial functions begin to fail and zombie cells start to accumulate, what happens?

Some cells in the body are not specialised, instead having the capacity to turn into many different types of cells. These are called stem cells. Stem cell exhaustion reduces an organism's capacity to repair and renew itself.

With all these things going wrong in the body, such as harmful chemicals being leaked by dodgy mitochondria and senescent cells, the body's immune system kicks into overdrive, which results in chronic inflammation or 'inflammageing'. This results in a weaker immune system as well as things like osteoarthritis, and is a risk factor for frailty, depression, dementia, kidney disease and more.[4]

That is a lot! The hallmarks of ageing are very comprehensive. But they do have their critics. What qualifies as a hallmark is a little bit arbitrary, and in 2023 the hallmarks were expanded to include three more categories.[5] There is also confusion around what is a cause, and what is an effect.

Is ageing a disease?

Maybe things could be simpler than the hallmarks portrait of ageing? Ageing should be considered a disease, plain and simple, according to David Sinclair of the University of New South Wales and Harvard Medical School. In his 2019 book *Lifespan: Why we age – and why we don't have to*, he argues that we have been lulled into thinking ageing is inevitable because it is ubiquitous. Most diseases are age-related, yet medical research focuses on treating the disease, not the ageing that made it more likely. Sinclair aims to eliminate ageing completely.

Why does he think this is possible? He has his own hypothesis, the information theory of ageing, which is subtly different to the hallmarks of ageing. He lays the blame for ageing at the feet of the epigenome, the 'filing system' that allows cells to know which bits of DNA to read to make proteins and when. This filing system can be mixed up when DNA is damaged and repaired. For instance, if you get sunburned, some of your DNA might be damaged. If the skin is fixed up, the DNA code will likely be the same, but the epigenetic markers attached to it might be a little wonky.

What does this have to do with stopping ageing? As the major damage is not to the genetic code itself, this means there is no permanent damage to the body's instructions, just that the filing system needs to be fixed. The body does have processes to tidy our genetic filing systems, and Sinclair believes therapies that enhance this process might end ageing.[6] This is a very dramatic claim and if he is right, it would have huge consequences – moral, religious, scientific, and legal. Sinclair is a very convincing communicator, but there are many researchers who think his ideas, while compelling, don't do enough to explain, and therefore eliminate, ageing.

How we gained then lost our immortality

We have a contested but workable understanding of what ageing is. So how did it come to be? Maël Lemoine of the University of Bordeaux is a philosopher of science and medicine and is particularly drawn to ideas around disease and evolution. The hallmarks of ageing, even with their limitations, are a great way to consider different aspects of the process independently. Lemoine was driven to create a history of ageing and investigate at what stage each hallmark appeared. Many species alive today are similar to our distant ancestors, like single-celled organisms. Lemoine built his timeline of ageing by synthesising pre-existing research into ageing in modern analogues of our long-lost ancestors.[7]

Even simple single-celled organisms like bacteria age. This is kind of weird when you think about it. Prokaryotes, such as bacteria, reproduce by copying their DNA, creating two almost identical daughter cells from a mother cell. This creates a little paradox when thinking about ageing. If a mother cell ages, when it splits it would pass on that ageing to its daughters. They would then continue to age until they split, passing on their additional ageing. The cycle would continue down the generations until the species was so aged it died. Bacteria have been around for billions of years, so clearly this has not happened. What's their secret?

Over time a single bacterium accumulates dodgy misshapen proteins. This is the loss of proteostasis hallmark, one of the genetic causes of ageing. To get around this, bacterial reproduction is not a pure 50/50 split. One daughter is youthful and 'rejuvenated' and the other inherits ageing. By a chance of chemistry, 'aged daughters' inherit misshapen and dodgy proteins from their mother. This means one line of daughters becomes increasingly aged until they can no longer function, while the others enjoy perpetual youth. Lemoine says the problem of dodgy proteins building up is 'universal', so all species must find a way to deal with it.

Eukaryotes are more complex than prokaryotes like bacteria. These are the organisms whose cells have energy-releasing mitochondria and nucleuses packed with DNA. They can be unicellular or multicellular, like us. Single-celled eukaryotes deal with ageing in a similar way to prokaryotes, creating 'rejuvenated' and 'aged' daughters. However, many more of the hallmarks of ageing popped up in the transition from prokaryote to eukaryote. With more DNA comes more information and information management. Hallmarks that relate to how DNA is read make their appearance here, as well as the nutrient-signalling hallmark. Many eukaryotes are adaptable, which is a great strength, but Lemoine believes these hallmarks of ageing are a 'side effect' of that adaptability.

Simple cells age, complex cells age. But Lemoine tells me that early multicellular animals might have cheated the system, saying, 'We have enough evidence to seriously hypothesise that they were immortal', so long as nothing ate, crushed, or otherwise destroyed them. It seems like a wacky claim, but the animals on Earth today most like these early ancestors might also be immortal. Sponges can live for thousands of years. Many researchers believe some species in the stinging group of animals called cnidaria, which includes corals, jellyfish and sea anemones, are truly immortal.

How did early multicellularity result in immortality? The secret lies in their simplicity. Today's cnidarians and sponges are not perfect analogues of primordial animals – they have had a lot of time to evolve. But they share the trait of having a relatively simple body that is easy to build. If you break some sponges apart, put them through a sieve and leave the mixed-up cellular soup in water, the cells will find each other and reform. Amazingly, the tiny freshwater cnidarian *Hydra vulgaris* can do the same thing. If you cut a piece off a hydra, the disembodied lump will grow into a whole new animal.

This is because most of the cells in 'immortals' can divide and build a new organism from scratch. In this way Lemoine believes they are 'just like a colony of unicellular organisms [that] could in principle renew endlessly' – just as in a cell colony 'all individual cells are doomed to age', but lines of 'rejuvenated' daughter cells can go on to live forever. The difference between cnidarians and colonies of single-celled creatures is that the cnidarians are a bit more organised.

Cnidarians and sponges take minimalist living to extremes. Sponges have no head or digestive system or nervous system, and filter-feed from their environment. Cnidarians have simple digestive systems, pooing out the same hole they eat with, nerves, but no brain or head.

For other animals, things are more complicated. Even the simplest worm has a front end and a back end. We have heads, internal organs and, if your draw a line down the middle of us from head to bum, we mirror ourselves. The symmetry we share with all animals that are not cnidaria or sponges gives us our name, bilaterians.[8]

Our bilateral body plan was a massive evolutionary advantage. Suddenly we could travel efficiently, becoming better hunters and escape artists, and giving us the capacity to move to where the mates were. But our sexy symmetrical bodies with brains, digestive stems and other organs came at a price: ageing. Most of the cells in our bodies are constantly being replaced. The cells in our lower colon get turned over every few days and the cell population of the heart gets replaced in just under five years. Most brain cells we take from cradle to grave, but recent research has shown we can gain new brain cells in adulthood. However, Lemoine explains that in complex bodies cells cannot just be replaced and rejuvenated endlessly as in some cnidaria, meaning we cannot 'rejuvenate' ourselves into perpetual youth.

Although some exceedingly simple worms that do not have many organs are candidates for immortality, replacing their bodies in a similar way to cnidaria, Lemoine hypothesises that the shift to a bilateral body plan, with all its avenues for complexity, is when ageing snuck back in. Bilaterians were now impacted by the hallmarks of ageing that occurred in prokaryotes and eukaryotes and they added their own hallmarks. These related to the communication between cells. As some of us complicated animals live a long time, hallmarks that relate directly to mutations on DNA in body cells and telomere attrition also appeared.

So why would bilaterians make this sacrifice? Early animals might have been immortal in theory, but they were not in practice because of predators and other dangers. Lemoine explains, 'They probably never use that [immortality], whereas you could use extra speed or agility to escape from your predators.' Perversely, giving up immortality could improve lifespan. By evolution's rules of surviving and reproducing, the bilaterians gained more than they lost.

Despite being lumped with ageing, many complex animals managed to evolve longer life spans. A 2020 study showed how specific genes that seem to promote longevity in mammals are related to the cell cycle of DNA repair and cell death, which maps onto the hallmarks of growth and metabolism. In long-lived species, these genes showed less diversity than other genes in that species. This implies that when a gene related to longevity in a long-lived species undergoes a mutation that undermines its function, that mutation is heavily selected against. This reveals that once longevity evolves, natural selection can help keep it in place.[9]

Why haven't we evolved to live forever?

It has been more than half a billion years since the first animals with heads are thought to have evolved. In all that time, why have we not evolved out of ageing? As Sadie the schnauzer and I illustrate, life spans are incredibly variable across the tree of life. This is a clue that the circumstances a species finds itself in create specific selection pressures that affect longevity. If we zoom out and compare animals with similar life spans across the tree of life, intriguing patterns emerge.

Large animals tend to live longer than small ones. Many whales become centenarians, and bowhead whales are thought to be able to make it to 200 years old. Elephants often plod on to 70, but small land-dwelling mammals like mice or tiny marsupials are lucky to see more than a year.[10] Across species, the riskier an animal's life, the faster it will age, and being a small mammal is a risky business. Why?

Twentieth-century British biologist Peter Medawar noticed the tendency for species who live a nice relaxing risk-free life to be long-lived. He got thinking about this just after evolutionary biology incorporated genetics in the 1940s – the moment in science called the 'modern synthesis', when evolutionary biologists were newly focused on how natural selection operated on genes. Medawar reasoned that it wasn't the stress-free lifestyle of these lucky creatures causing them to age slowly, instead, risk must somehow influence how genes are, or are not, being selected for.

Over generations little changes – mutations – randomly occur in our DNA. Sometimes these changes are beneficial and might be selected for. Most of the time mutations are harmful, reducing an animal's capacity to thrive in its environment and even leading to disease. Usually, natural selection weeds out these harmful mutations as individuals who possess them either don't breed or breed less than their peers.

Ageing often accelerates after reproduction. Frazzled new parents might be nodding along to this, feeling the demands of their kids have aged them. But sleep deprivation or stress was not what Medawar had in mind. He thought if the mutations that cause ageing only come into effect after reproduction, they would be less visible to natural selection. In most species, once the next generation is born, hatched or, in some cases, raised, their survival is completely independent of parental health. So mutations that rear their ugly heads after this point can be passed down. In that way young parents hand down their future ageing to their kids. This is thought to be one reason cancer rates start to steadily increase in humans after we turn 40.

Humans tend to live long past the age of reproduction, when our capacity to do so ceases or reduces due to menopause and lowered sperm quality. Some animals, like octopus, salmon, and many small Australian marsupials, breed once and then die. But many other species can keep breeding year after year. What about them? Surely the more years they live, the more offspring they will have, and selection would favour mutations that increase life span?

This is where the danger comes in. Even in species that can breed indefinitely, death by disease, accident or predation is always prowling nearby. So every individual in a population will stop reproducing eventually, because it is dead. Therefore ageing mutations cannot be weeded out at this point – this is called a 'selection shadow'. In 1951 Medawar launched this idea in a lecture titled 'An Unsolved Problem of Biology' and published it in a paper a year later.[11] A quick survey of animal life does seem to back Medawar up.

There is a strong correlation between being able to fly and slow rates of ageing. Flight makes life less risky. The environment gets dicey? Take to the wing! A predator is chasing you? Fly away! The Brandt's bat, which at 7 grams weighs about as much as a pencil, can live for over 40 years. Another way to avoid danger is to head underground. Burrowing mammals age more slowly than their surface-scuttling counterparts. Being poisonous, spiny, or having poisoned spines all seem to be tickets to longevity.

If you wish humans could live longer, we can use Medawar's logic to blame dinosaurs for our life spans. Famously, many reptiles live for a long time. And as Fred the 110-year-old cockatoo demonstrates, today's feathered dinosaurs, the birds, are often long-lived. In other words, mammals have strangely short life spans. Most mammals during the time of the dinosaurs were small, insectivorous, and nocturnal. This has led João Pedro de Magalhães of the University of Birmingham to suggest that as life would have been dangerous for our fluffy ancestors living at the feet of dinos, mammals were most likely short-lived until the extinction of the non-avian dinosaurs, when some of them could evolve to grow larger and enjoy longer lives.

In 2023 De Magalhães suggested that the 165 million years mammals spent scuttling through the night during the reign of the dinosaurs put us through a 'longevity bottleneck'. Our life spans were compressed in a way that did not occur for other land vertebrates like birds and turtles.[12] He writes: 'Such a long evolutionary pressure on early mammals for rapid reproduction led to the loss or inactivation of genes and pathways associated with long life.' We survived the dinos by pumping out babies quickly, and were preyed upon before we could pass on any longevity genes, leading to their deterioration.

Selection shadows, like those described by Medawar in the 1950s, are also being observed in modern animals. These shadows shorten the longevity of a species swimming around in Africa today. Pretty little killifish live in diverse freshwater systems. There are many species of killifish, and some live in nice lush waterways that the fish can enjoy all year round. The species that live in these cushy conditions can live for many

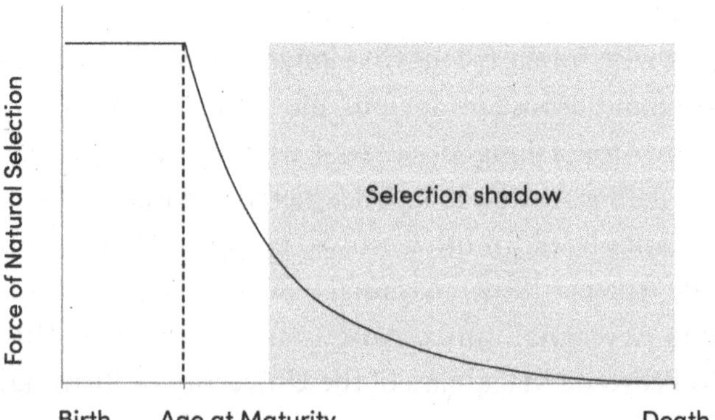

years. Other killifish are harsh environment specialists, making the best of it in freshwater bodies that regularly dry up. When the water goes, the adults die but, fascinatingly, embryos can avoid being desiccated by the sun for years, meaning that when water returns, they can start their lives. As Medawar would predict, the harsh environment specialists age more rapidly than their cushy counterparts.

The long-lived and short-lived killifish are very different from each other genetically. This was shown in a 2019 study led by Rongfeng Cui of Sun Yat-sen University and colleagues. They showed that the fish from the harsh environment had way more DNA in their chromosomes compared to the cushy killifish. This was due to a build-up of mutations and DNA repeats. Genes associated with ageing were particularly likely to have mutations, suggesting that the certain death these fish faced when their ponds dried up limited natural selection's capacity to weed out these mutations. When the researchers looked for a similar pattern in humans, they found it, but to a lesser extent. In humans, genes related to longevity were more likely to be mutated than other genes in the genome.

But there is a spanner in the works: a dangerous life does not always select for accelerated ageing. Despite the elegance of Medawar's thinking, there are many limitations to his big idea, and the 'problem' remains 'unsolved'. However, the possible link between risk and ageing does seem to be scattered throughout the animal world, suggesting it is a piece in the larger puzzle.

A brutal but scientifically brilliant experiment on nematode worms showed that in specific circumstances, longevity can evolve in the face of danger. Hwei-yen Chen of Lund University created two populations of worms. Both groups were subjected to high death rates. One group's causes of mortality mimicked random causes of death. The other group was exposed to 40-degree C temperatures for extended periods. Survivors from each group were then allowed to breed within their group.[13] Just as Medawar predicted, the population of worms subjected to random deaths evolved shorter life spans. But, contrary to his logic, the worms that faced the heat generation after generation and survived had longer lives and genes that corresponded to longevity.

The researchers suggested 'robust' worms that survived the heat may also have had genes that made them longer lived, so those genes became more common in the population. The takeaway? When life is full of many dangers, a shorter life may evolve, but specific dangers might lead to increased life spans through natural selection.[14]

Medawar thought ageing in the wild would be exceedingly rare, as wild animals did not live long enough to start the ageing process. This would mean genes that led to ageing were invisible to natural selection. The idea that wild animals seldom experience many of the symptoms of ageing is the premise his work rests on, but it is not true. Wild animals have been shown to lose condition, strength and reproductive capacity with age. And, like humans, as they age their chance of dying from all causes increases. In other words, animals do age in nature. Despite the evidence, the idea that wild animals do not age remains a persistent myth.[15]

Even incredibly short-lived animals age. I would not like to be an antler fly; factors such as predation kill about 13 per cent of them each day. Oof. Because of this, male antler flies live an average of six days. However, lucky ones may live for over a month, and during that month they age. This was shown by Russell Bonduriansky, keeper of the stick insects who I met earlier at the University of New South Wales.

The tiny antler flies live their lives on discarded moose antlers. Early in his career Bonduriansky studied sexual selection in this species and minutely tracked their reproductive lives.[16] Because the flies live and die on the antlers, he was able to mark individuals with a unique identifier and follow their entire lives. After a gruelling 72 days of fieldwork in the forest observing 609 male flies every two hours, Bonduriansky had an amazing dataset. He had risked a lot to do it, having to set elaborate bear detection systems that would blast music when bears came near. The data was collected to study the flies' sex lives, but Bonduriansky realised that it also described their ageing – something that had never been shown in wild insects. The numbers showed that with each day that passed a male mated less and had an increased risk of dying.

When antler flies and many other wild species start to age, they become more vulnerable to death by other means. So while ageing is common, is being elderly also common? Do wild animals ever die of old age? Do they become truly frail, as humans and domestic animals do? It is rare, but can happen. In 2013 news broke that a female brown bear, known as No. 51, was not doing well. Researchers had fitted her with a radio collar in 1981 when she was seven years old. She lived in the forests of Minnesota and became a favourite among researchers. Karen Noyce, a scientist at the Department of Natural Resources, made a statement asking shooters to leave the elderly animal in peace. 'She doesn't hear much and can't see much … Her gait is a little unsteady. When people see her, they think she looks drunk. That's because no one ever sees an old, old, old bear like that. But she doesn't seem to be in any pain.'[17] Later that year No. 51 died at the age of 39 and a half, her cause of death assumed to be old age.[18]

Can evolution 'select' for ageing?

Ageing is common in nature, even if becoming truly elderly like No. 51 is rare. This implies there must be more going on than a selection shadow. What if the cause of ageing was a little more direct? What if genes that cause ageing could be selected for? It's a weird thought: how could natural selection actually select anything that causes such a raft of problems? It could be possible if those same genes provided benefits early on. This is the idea behind another big hypothesis of ageing, which has the tricky-to-pronounce and very jargony name of antagonistic pleiotropy. Pleiotropy is the word for when a gene can cause two or more different traits.

This idea also has a long history, having been launched by George Williams in 1957. This is the same Williams who coined the term 'gene's eye view' discussed earlier in this book. Antagonistic pleiotropy extended the idea of a selection shadow, but Williams noted that the impacts of ageing are felt in many species before true old age sets in and during reproductive years, when a fair chunk of the population can still expect to be alive. This could be the result of direct selection for a gene that caused ageing if it was also helpful, increasing the chance of reproduction, in early life. He writes:

> An advantage during the period of maximum reproductive probability would increase the total reproductive probability more than a proportionately similar disadvantage later on would decrease it. So natural selection will frequently maximize vigour in youth at the expense of vigour later on and thereby produce a declining vigour (senescence) during adult life.[19]

The original paper describing antagonistic pleiotropy was very speculative. The idea was great, but where was the evidence? Even Williams expressed his frustration, writing that 'convincing examples are hard to find'.

Genetic technology has improved a lot since the 1950s, and evidence that ageing is selected for in this way is becoming more solid. Evolutionary biologist and ageing researcher Jessica Hoffman explains to me that 'A lot of our genes that we know are bad for ageing... are really good early in life.' Hoffman, who researches at Augusta University in the United States, is trying to nail down the specifics of exactly how these genes work and the processes they are involved with.

I gravitated to Hoffman's work because of her unusual study subjects: dogs. Small dogs like Sadie often live to 15, and sometimes will even make it to 20, but giants like Irish wolfhounds and Great Danes are old at seven or eight. This is due to a quirk of biology where within a species larger animals tend to have a shorter life span, even though the reverse is generally true between species. This is the case in humans, where short people tend to outlive tall people.[20] The trend is far more evident for dogs, as they vary in size so dramatically. When Hoffman first started working on dogs, she was driven to ask, 'Why is it across species being bigger is better, but within a species being smaller is better?'

The answer: antagonistic pleiotropy. At least in part. Larger dog breeds have bigger litters. Hoffman explains that genes that are 'good for growth and development' are often bad news in later life. But, if those genes also result in more young, 'it makes sense that they would be selected for'. As Williams suspected, reproductive success is being traded off against life span. Dogs are strange because humans have bred them for specific traits, but their artificially selected diversity of size exaggerates some of the processes at play in antagonistic pleiotropy. Specifically, Hoffman points the finger at growth hormone as a driver of higher reproduction but shorter life span.

Life span and reproduction might be traded off against each other within our species. Studying longevity in humans is notoriously complicated. We live so long that a whole human life span is hard to study. Things are made extra messy as our cultures, diets, socio-economic status and sex all play a part in our ageing. But early onset of puberty – an early investment in reproduction – has been shown to reduce life expectancy in multiple studies of cisgender men and women. Puberty onset and life span are both multifactorial, but one of those factors is genetics. And the connection remains clear even when factors like diet are taken into account.

Frustratingly, this research has not been put into an evolutionary context. In a massive 2015 study of half a million people, neither evolution nor antagonistic pleiotropy were mentioned.[21] In a 2020 study investigating the genetic drivers of early puberty in boys and life span, pleiotropic genes were discussed, but evolution was not. However, evolutionary biologists have found this type of research and used it to speculate about antagonistic pleiotropy.[22]

Very little work has been done to show this mechanism in the wild. But remember those killifish and their long-lived species living in stable environments and rapid agers living in pools prone to drying up? While the short-lived fish had way more damaging mutations in their immune system compared to the old souls, the trend was the opposite for genes related to development and reproduction. In this case the researchers stressed that while antagonistic pleiotropy might be at play in the rapid ageing of killifish, the accumulation of harmful mutations over time was likely a larger force. Hoffman believes antagonistic pleiotropy is a key driver of the evolution of ageing across animals, but she agrees there is more to it, saying, 'I'm not convinced it is the only driver.'

How else could ageing be selected for? In the 1970s, Thomas Kirkwood came out with a third big hypothesis to explain the evolution of ageing – the disposable soma theory.[23] Kirkwood is now an emeritus professor at Newcastle University's Institute for Ageing in the UK. Soma is another world for body, and the hypothesis essentially views the body as a disposable tool for reproduction, an idea similar to Richard Dawkins' view of bodies as 'lumbering robots' functioning only for the survival of genes.

The idea hinges on the notion that organisms only have a limited budget of energy and resources to devote to different things like growth, reproduction, and DNA repair. With a limited budget, we have to cut back on spending, and ageing is a result of resources being focused on reproduction or growth at the expense of bodily upkeep. Again, it's a fascinating idea, but it has limitations. For instance, calorie restriction seems to be a sure-fire way to increase life spans across many species. If ageing is the result of a limited energy budget, why would limiting energy input result in a longer life?

A more controversial idea suggests that ageing could be selected for as it increases the 'evolvability' of a species.[24] In changing environments, short generation times mean that selection can be more specific. You can see how this might operate under climate change. If a 70-year-old sea turtle lays eggs, she is passing on great genes for being a baby sea turtle 70 years ago when the ocean was cooler. On the other hand, small marsupials called antechinus breed at 11 months old. After one breeding season all males and most females die, although some females will breed for two seasons. In this case animals that survive to breed will be well suited to their environment, taking seasonal variations into account, and as their joeys will be adults the following year, they will likely be very well adapted to the current environment. So shorter life spans might be selected for in species in changing environments where evolvability is at a premium.

Bringing up 'evolvability' as a possible driver of life spans might make some evolutionary biologists squirm. It evokes the idea that we 'age for the good of the species and to make way for the next generation'. This might be a comforting way to frame ageing, but it's not how natural selection works, there is no goal in evolution. We saw this with altruistic behaviour – it evolved when it popped up through random chance, provided a benefit, then was selected for. Altruism was not selected for because it would be good for the species, instead the good it was doing meant it was selected for. In the same way ageing has not evolved to be a gracious handing over of the baton from one generation to the next. However, if your ageing meant that your offspring and then their offspring had a better chance of survival, it could evolve. Or if groups of animals that aged faster had a greater evolvability than slow-senescing equivalents, it might happen.

Yes, but why is my dog ageing faster than me?

The ideas in this chapter are the heavy hitters when it comes to explanations for the evolution of life spans. There are other, less accepted, ideas, but like the hypotheses above, they all have their limitations. In both humans and dogs we are a long way from understanding exactly how our life spans evolved. Even so, many evolutionary biologists believe that by combining hypotheses old and new, we can go some way to explaining why we live as long as we do.

Both humans and dogs have increased our life spans along our evolutionary story. As far as mammals go, we are not very closely related. Our evolutionary paths diverged during the time of the dinosaurs, keeping us small and short lived for millions of years.

Dogs were domesticated from wolves when humans artificially selected them for traits like friendliness and cooperation with humans. This happened somewhere between 15 000 to 23 000 years ago, although some suggest the leap was made closer to 40 000 years ago. Wolves can live to age 13 in the wild, but usually die years before their tenth birthday. Hoffman believes that as no one has ever bred dogs specifically for life span, and as wolves have similar life spans to large dogs, dogs' rate of ageing has been fairly unchanged, except in the case of small dogs, who by virtue of their size can live longer.

Humans are primates. For mammals, primates are a long-lived bunch. Even pocket-sized primates can live into their twenties. Macaques, which are the classic 'monkey-type monkeys' can live to 40, though this is rare; meanwhile chimpanzees at Ngogo in Uganda can reach 60.

How does this stack up against the evolution of ageing hypotheses? Through time most species of primates have been tree dwellers. There is evidence to suggest predators are less likely to kill animals that live in trees compared to ground dwellers. This might have made life less dangerous, in turn increasing natural selection's power to weed out mutations of longevity genes. Primates tend to have only one baby at a time, although twins are more common is some species. Arboreal life may have also limited the number of young a primate had, as wrangling multiple babies is a deadly affair in the treetops.

We primates are also slow reproducers, reaching reproductive age at a leisurely pace, and thus individuals who had longevity traits may have had more time for successful reproduction, passing those genes on.

Wolves mature faster, but a 40-year study of Scandinavian wolves found that the age at which they first reproduced was highly variable, ranging from one year to eight, with an average of three.[25] Rather than having one bub like your average primate, they can have up to ten pups in a litter, though this is rare. Litter sizes can vary depending on conditions, however, litters of more than five are common.[26]

Humans are incredibly late breeders, even for primates. It takes a lot of time and investment to raise our offspring, and multiple births are rare. When compared to wolves and dogs, it seems likely that our differences in life span are at least in part the result of a classic trade-off between reproduction and longevity, whereas dogs breed far earlier and have more young at once.

Get old with a little help from our friends

Humans are weird. When you compare us to other mammals, we are freakish on many fronts. We do not have much hair, we can thrive in an incredibly diverse range of environments, and we have language. Another weird thing about us that often slips under the radar is that for our size, we are long lived. We outlive all other primates and reach similar ages to elephants, often outliving them.

Compared to other species, humans are social butterflies. And it turns out that social species have a longevity advantage over solo operators. This was demonstrated in a co-authored 2023 study led by Pingfen Zhu of the Institute of Zoology in the Chinese Academy of Sciences, using data on 1000 mammal species.[27] This is a little weird given that, as the Covid-19 pandemic has reminded us, social connection brings with it infection risk. This very risk puts a strong selection pressure on populations to evolve robust immune systems, and the researchers suggest our supercharged immune systems, forged by sociality, might increase longevity.

The other big danger of living socially is other people. Conflict and competition can result in violence or resource monopolies. Even so, the evidence is in that within many species the benefits of social connection outweigh the risks, and so we live longer with a little help from our friends. Humans with stronger social bonds live longer. The same trend has been found in female chacma baboons and rhesus macaques, demonstrating that in some cases there are serious benefits to a social life. This led Zhu and colleagues to suggest that living with others helps insulate animals from the dangers in life. This can be in a very basic 'safety in numbers' way, but also through collaborative behaviours and the sharing of knowledge – for example, the way sperm whales and long-finned pilot whales provide childcare for each other. Maybe our hyper-sociality has insulated us from some of life's dangers, driving our life span up.

But hang on, what even is our life span? Turns out, determining the human life span is tricky and controversial – who would have thought? Between 2012 and 2022, 31 Australians died at the age of 110 or older. Some researchers believe we are only a few medical tweaks away from getting people to 125.[28] In his book *Lifespan*, David Sinclair audaciously suggests that soon there will be no limit to the human life span. Many in the scientific community doubt this claim, including myself, but he does back it up with intriguing research. The reality is that in wealthy countries most people are considered to have lived a good long life if they make it into their 80s.

Have we always lived this long? A common myth about ageing is that until recently nobody lived past 30 or 40. I have seen and heard this notion promulgated in film, on social media, in conversations and, perhaps most shockingly, in a paper published in the prestigious peer-reviewed scientific journal *Nature Communications* in 2024.[29]

The myth is partly driven by a simple mathematical misunderstanding. Until recently child mortality was incredibly high. This is still the case when people do not get access to medicine and sanitation. If you average out everybody's age at death, and if many of those people died in their first few years of life, then the 'average age' is going to be low. But this does not mean people did not get old in the past, although getting old was a rarer occurrence.

In ancient Rome, the elite were able to age, and this is reflected in the age requirements for political office. In the Roman Republic, you had to reach 42 years of age before being eligible for election to the prestigious role of consul. Later, during the Roman Empire, many emperors lived long lives. Gordian I was thought to be 80 when he became emperor, and might have lived to be far older had he not ended his own life after a bereavement. Life for poor people might have been a lot harder, and shorter, than it was for the elites. Researchers have tried to figure out if many underprivileged ancient Romans got old. So far, the evidence looks grim, with life expectancy for those who survived childhood possibly being within the 30 to 40 range.[30]

Staying in the vicinity of Rome, a fascinating study comparing the life span of Vatican artists and popes provides an intriguing insight into the life expectancy of the well-to-do from 1200 to 1900. The study found the average life span of popes was higher than that of artists. From 1200 to 1599, popes died at an average age of 66, with the oldest pope reaching 72. On average artists kicked the paint tin at 63, with the oldest reaching 71. From 1600 to 1900 the average age at death for popes was 77, and for artists it was 70, with the oldest living to 82 and 79 respectively.[31]

Research into life spans in modern hunter-gatherer societies shows that while it is true fewer people reach old age, it is common for people to stay hale and healthy into their 70s. Of course, modern hunter-gatherers are not windows into the past, they are modern people living in specific contexts. In fact, different hunter-gatherer groups vary in how many of their members live to what age, showing that even in today's world, humans are so diverse there is no one way to understand how environment, genetics, lifestyle, and medicine interact with longevity.[32]

We *Homo sapiens* seem to have longer life spans compared to the earlier *Homo* species we evolved from. Today we also have longer life spans than the Neanderthals, *Homo neanderthalensis*, our cousin species that we did not evolve from but did interbreed with a little. But when exactly we started to live into old age is unclear. In the mid-2000s Rachel Caspari, an anthropologist from Central Michigan University, launched a series of papers suggesting that we only started regularly living past 30 years of age around 30 000 years ago. That is over 170 000 years after *H. sapiens* first evolved – perhaps nearly 300 000 years, depending which scientist you talk to.[33]

Homo Sapiens and Neanderthals were both living in Asia 100 000 to 40 000 years ago, in a time called the Middle Palaeolithic. Caspari and Sang-Hee Lee analysed the teeth of Neanderthals and *H. sapiens* adults from this time and categorised them as either 'young' or 'old'. This distinction was not based on chronological age exactly, but as teeth 'erupt' at different times in early adulthood and wear in predictable ways, they were able to estimate when adults were double the age that they could first have babies. At this age they were considered 'old', as they could be grandparents. For *H. sapiens*, this age was around 30.

Asian Neanderthals and the *H. sapiens* at this time had very similar ratios of old to young people. Nine Neanderthals were 'old' and nine were 'young'; six *H. sapiens* were 'old' and eight were 'young'. Interestingly, the European Neanderthals at the same time were far less likely to become old, with the sample showing 36 'old' and 103 'young'. This suggests how climate and culture might have impacted the life span of our cousin species.

The researchers then zoomed forward in time to look at more recent *sapiens* samples. These were from around 30 000 and 20 000 years ago in a period called the Upper Palaeolithic. Caspari and Lee found the ratio between old and young in *H. sapiens* adults had dramatically changed. Far more people were reaching 'old': 88 had lived to die in the 'old' category and 42 died when they were 'young'.[34] These samples were from Europe, where the climate was harsher than it had been in Middle Palaeolithic Asia, making this longevity even more remarkable.

Caspari has argued that her research might reveal that living grandparents first appeared on the scene in the Upper Palaeolithic, many tens of thousands of years after we first evolved.[35] I think her argument is fascinating, but I do not believe there is enough evidence to support it, though that may come. The sample sizes, especially from the Middle Palaeolithic, are very small, so might not be great representations of the demographics of the time. Additionally, humans were relatively new to west Asia during the period she was sampling. I would love to know how their life spans compared to people in other parts of the world, specifically Africa. Modern humans had been in Africa for tens of thousands of years at this time. To me this suggests they might have been more physiologically and culturally adapted to their environment and may have regularly made it to old age.

However, if humans only started making it to old age 30 000 to 20 000 years ago, this has tantalising implications. The Upper Palaeolithic seems to be a time of huge cultural shifts. Archaeological evidence suggests that across Europe, Asia and Africa new technologies like grinding stones were invented, and art and jewellery became more common. Caspari suggests grandparents could have been behind this shift. If more people were living past 30, there would have been more elders in the community. As soon as elders started to become more common, this could have created a feedback loop whereby the wisdom, cultural stability, knowledge of changing environmental conditions and economic contributions of the elders made it possible for more people to survive to become grandparents. This in turn could have fuelled the cultural revolution.

Caspari's work is an intriguing thought experiment, but it is so contingent on data from Eurasia I am wary of drawing a universal conclusion. However, I am intrigued by the questions she is asking, even if I'm not sold on her conclusions. Knowing if there have always been elders in our communities is critical if we are to understand our social and evolutionary history. This is because grandparents, especially grandmothers, are of particular interest to evolutionary biologists.

Thanks to our elders

Want to start a fight on social media? When scientific publications post about menopause, they usually go with headlines like 'New insights into why women live so long after reproducing'. Inevitably the comment sections are filled with rage along the lines of 'Why does science find it so hard to see value in women outside of their capacity to reproduce?' and 'Women are not just babymakers'. I agree politically, but biologically it makes us freakish, and scientists win no friends when they try to explain the in impenetrable academic language. So let's disentangle the topic and understand how these social media misunderstandings occur.

Menopause is a rolled-gold, rare VIP trait we should be proud to share with a selected few, including orcas, beluga whales, short-finned pilot whales and narwhals. In 2023 it was announced that chimpanzees are also in the menopause club.

A famous, yet still hotly debated, argument for life after menopause is the grandmother hypothesis. This suggests that when a female's genetic fitness is increased by helping raise grandchildren, menopause can occur. The idea was first toyed with by George Williams in his famous 1957 paper, but has since been championed by Kristen Hawke of Utah University.[36]

Orcas go through menopause in their late 30s to early 40s but often live into their 60s. One famous orca elder named Granny was estimated to live to 105. Matriarchs like Granny are central to orca sociality as pods are matrilineal. Daughters don't leave, and the oldest female provides wisdom and guidance to the group. Research has shown that killer whale grandmothers increase the survival of their grand-calves. Crucially, this effect gets stronger when she is no longer reproductive.

When times are hard, the wisdom of these finned ladies becomes particularly precious to the pod.[37] This has been demonstrated in salmon-eating pods, where the death of a grandmother increases the mortality risk of her grand-calves in any year, but that risk is greater in years with low salmon abundance. As the impact of losing a grandmother orca is more severe after her menopause, whales that go through menopause have more descendants because of their wonderful grandparenting, who inherit the genes for 'the change'.

It is possible similar dynamics are at play in narwhals, belugas, short-finned pilot whales and humans. Kristen Hawke provided strong evidence for this in her work with the Hadza people in the 1980s and 1990s. The Hadza are a modern foraging and hunting people from Tanzania. Most of the calories eaten by this community are gathered by women. While mothers and children forage, grandmothers focus on procuring labour-intensive foods like large tubers. Their contributions become particularly valuable when their daughters have newborns.[38] As Hawke laid it out in an interview with *Smithsonian* magazine, older females were 'subsidising the fertility of younger ones'.[39]

Beyond the Hadza, this pattern has been found in humans across various times and cultures. The Finns are great record keepers and an analysis of demographic data from 1761 to 1900 showed that the presence of maternal grandmothers protected their grandkids against all causes of mortality, and specifically that having your mum's mum about meant you were less likely to die of an infectious disease as a child. The effect was not found for paternal grandmas.

Church records detailing the lives of French-origin families in pre-industrial Quebec give a fascinating peek into how grandmothers helped in this challenging 17th- and

18th-century environment. A woman whose mum was alive when she had kids had more kids, and more of those kids lived to at least age 15, compared to women who did not have a living mother. This was not a result of inheriting genes that were suited to the environment, as when grandmothers were alive but lived far away the effect was not detected. This showed it was the presence of the grandmother that provided the survival boost, not her genes. This study did not look at the impact of paternal grandmothers.

All this data seems like a slam dunk for the grandmother hypothesis. But things might not be as straightforward as they seem. Humans also tend to move about, with women leaving communities to marry, decreasing their proximity to maternal support when raising the kids, so it is possible that while grandmothers can aid survival, they have not had the chance to do so enough for their contributions to be significant in terms of evolution.

While it seems there is good evidence for menopause in an orca grandmother upping the survival odds of her grand-calves, elephant grandmothers seem to have evolved a way of having their cake and eating it too. They also increase the survival of their grand-calves, and while their fertility decreases in old age, it does not end. So maybe menopause is the result of a biological constraint – we just can't be infinite baby machines – rather than active selection.

The 2023 news that female chimpanzees, who stop reproducing at 50, can live into their 60s at relatively high rates, presented another challenge to the grandmother hypothesis.[40] Menopause in this species had been observed elsewhere, but females did not live long after it. These old girls at Ngogo where the study took place were showing researchers that the non-reproductive life of a chimp can span over a decade. It is possible that the pleasant conditions at Ngogo allowed them to live longer than had previously been observed. As female chimps tend to leave their community when they become adults, leaving their mothers, selection for menopause in this species through the effect of grandmothers seems less likely – although some females do stay in the community they were raised in and tend to be very close to their mums.

Before 2023 it was supposed that menopause evolved late in the human story, either in our species or in a not-too-distant ancestor. As we saw, Caspari suggested it only emerged around 30 000 years ago, which would mean that it evolved, or became apparent, in multiple disparate human communities independently. But chimpanzee menopause calls this into question. Maybe it is more ancient, dating back to our common ancestor? Or maybe we evolved it separately? If wisdom and grandparental care were not the driving forces selecting menopause in our ancestors, they still might have had a role in extending the length of our non-reproductive life span.

What about grandfathers? Sperm quality and reproductive capacity in humans decreases over the reproductive lifespan, with the slope significantly steepening from the late 40s. Reduced sperm quality does not completely prevent paternity in later life, but it puts the brakes on. Senescing sperm is also seen in laboratory rodents, which leads to the comfortable assumption that this is the way it goes across species. But a co-authored paper led by Krish Sanghvi of Oxford University published in *Nature Communications* in 2024, found that humans, lab rats and mice seem to be the odd ones out.[41] By analysing the results of 379 studies into 157 species, researchers found no consistent pattern of ejaculate traits over life span.

This truly groundbreaking, showing that selection is pushing life spans beyond the point many people with sperm stop reproducing. Unfortunately, the researchers fell into the 'life expectancy trap'. They argued – citing research that calculated average life expectancy from birth that included child mortality – that men only started to live beyond 'the time their sperm was maintained' within the last 200 years.[42] A warning to us all that errors can slip by even in prestigious, peer-reviewed publications such as *Nature Communications*. However, the logic – that reproductive capacity and life span are linked – could still be applied if human longevity increased beyond 40 in the Upper Palaeolithic, or before.

Is it time for the grandfather hypothesis? Not much work has been done calculating how grandpas impact the survival and abundance of their grandkids. Do grandfathers who live longer have more grandkids? The data here is mixed. In the Tsimane', a forager-horticulturalist people from Bolivia who live in small-scale communities, grandfathers provide significant material support in the form of food to grandkids, as do grandmothers. This is true into their 70s.[43]

Finnish data from 1761 and 1895 showed that having a living grandfather or grandmother increased the odds a grandkid would live past the age of one – but only if the grandparents did not live with their grandkids. Co-residing grandfathers actually decreased the chances the grandkids would survive, whereas a co-residing grandmother did not impact the youngsters' chances one way or the other. The researchers contended that when grandad lived with the family, he might compete for the resources necessary for the survival of the infants. To me, this research highlights how culturally specific many of the selection pressures that have shaped us have been.[44]

Are we missing something? Are we too focused on direct descendants? For most of human history we have lived in small communities with high levels of relatedness across the group.

As we have seen, humans can be sharing and collaborative even with people outside our immediate families. One of the ideas in Caspari's work suggests that the sudden explosion in culture during the Upper Palaeolithic was because having older people in the community helped increase the knowledge the community held.

Elephant family groups are led by a matriarch. Researchers have found that as matriarchs age they get better at identifying threats, such as lion roars, and thus the survival rate of the group increases with the leader's age.[45] Male elephants leave their families as adolescents, but it's not a lonely life. While their social relationships are more fluid than those of the females, bulls often travel together in groups. In these groups older males seem to make the decisions about where to go, which suggests elders are important even for this sex-segregated species.[46] When wolves fight over territory, one of the deciding factors that determines their success, other than pack size, is the presence of older adults.[47]

While humans likely have pushed our life spans up by being good grandparents and elders, it is possible that the positive effects of our senior ancestors in groups have been more expansive, allowing us to succeed in a range of environments with the benefit of decades of hard-won knowledge.[48]

CHAPTER 7

Why do we drink?

Alcohol is a poison. But it is a poison many of us love. And its allure is not confined to humans – many species in the wild have a taste for booze or can develop it in a lab. But not all animals can hold their liquor. Fruit bats can partake of alcoholic nectar and still manage to navigate the night sky, whereas elephants on the hard stuff soon start to stumble. This is because some species have evolved the capacity to metabolise alcohol, allowing them to consume it in small quantities without getting too drunk. Humans are one of those species. It's this capacity that hints at our love affair with the bottle – one that goes back in time, well before bottles were even invented.

For some, alcohol has caused nothing but pain and sorrow; it is a visible and ever-present villain. For others, in cultures where it is prohibited, it lurks as a shady, stigmatised substance, obtained through underground networks and consumed in secret. And of course, alcohol can be wonderful. It's part of religious rituals, it's drunk to mark birthdays, graduations, first dates and rites of passage. It is a hero. I have always found these contradictions hard to reconcile, personally, politically, and scientifically.

Alcohol's villainous side has caused pain and grief for population geneticist Ben Clites. Clites grew up in West Virginia, a part of the United States hit hard by addiction. He explained to me that 'the opioid crisis was really bad there.

I had a lot of friends that died, a lot of friends whose dads were vets and alcoholics, and [alcohol use disorder] runs in my own family'. Looking for answers, he turned to science and started a PhD at the Waggoner Center for Alcohol and Addiction Research at the University of Texas, Austin. At conferences, lectures, and seminars, experts would trot out data points on the danger of alcohol use and the rhythms of addiction, but to Clites something about these talks did not seem right.

Humans have a wariness of dangers in the environment, like snakes and heights.[1] Not so for the dangers of alcohol. We did not evolve a distaste for drinking; bizarrely, many of us actively like it. We imbibe even while understanding it is risky. Clites worried that his colleagues were missing something by not investigating the evolutionary reasons why humans like to drink.

A big reason we drink is because our brain rewards us for it. As alcohol enters our bloodstream and then our brain, it triggers the release of a particular set of neurotransmitters, brain chemicals that tell us 'well done'. These include endorphins, which are chemically like opioids, making us feel good, sometimes euphoric, and reducing our sensitivity to pain. The endorphin release boosts dopamine and serotonin. Dopamine is our brain's way of keeping us on task. It rewards us for striving and accomplishments and encourages us to repeat the action that triggered its release. Serotonin is associated with good moods, among other things. When addiction strikes, it is the release of these chemicals that people become addicted to.[2] Addiction itself has a broader evolutionary story that intersects with specific addictions.

Addictive drugs other than alcohol interact with our brain's reward system in direct ways. Opioids are a similar chemical structure to endorphins and thus produce similar effects in the brain. The active ingredient of cannabis, THC

(Tetrahydrocannabinol), impersonates chemicals in our bodies called cannabinoids, which in turn trigger a dopamine release.[3] Cocaine is more direct, blocking the system that clears away dopamine, increasing its build up.[4] Alcohol is different. It does not trigger our reward and pleasure systems by deception, imitating something the body already makes. Alcohol is a small 'promiscuous' molecule, Clites explains. It washes through the brain changing many things, but we like it because it triggers our own happy chemicals to be released.[5]

The release of happy brain chemicals explains both why we like alcohol, and how we can become addicted to it. In evolutionary biology this explanation is called a proximate cause. A proximate cause for why birds fly is that they have wings. But this was not enough for Clites, he suspected that to better understand humans' relationship with alcohol, researchers should focus on its 'ultimate cause' – the evolutionary history that led to a certain trait being selected. An ultimate cause for why birds fly might be to avoid predators, to travel long distances or to access food and nesting sites. He wanted to understand why our evolutionary history selected for a system that rewarded us for drinking.

Did our love for grog survive because it's not that bad for us after all? Sadly, this is not the case. It is a killer. Drinking makes us risk takers, increases the chance of accidents and is associated with violence. Chronic excessive alcohol use wreaks havoc on the body, changing the brain, causing liver disease and a host of other maladies. This is usually caused by alcohol use disorder, colloquially termed alcoholism.

Even in small quantities, alcohol can be dangerous. Drinking while pregnant can result in a child being born with foetal alcohol syndrome, which can cause significant difficulties throughout the baby's life. Moderate use increases the risk for several types of cancer, including breast and bowel cancers.[6] Drinking increases the risk of contracting the deadly bacterial infection tuberculosis. While tuberculosis is preventable and curable, inequitable access to treatment means it is still one of the globe's biggest killers.[7]

But what about red wine? A glass or two of red is one of my favourite things to enjoy while cooking or chatting with friends. It is often heralded as helping heart health, so I always thought it couldn't be too naughty. Its benefits have been backed up by many studies that correlate moderate drinking with reduced risk for cardiovascular disease and diabetes compared to abstainers or heavy drinkers. However, some studies find no positive effect at all, and many researchers are now urging caution before we open a bottle. These studies rely on self-reported data from large groups of people over time, so are open to inaccuracy. Additionally, disentangling cause and effect is tricky, as moderate red wine drinking is often enjoyed by people with high incomes, balanced diets and other traits that make them healthy. So it is probably too soon to say we evolved to like drinking because it was good for our hearts.[8]

But now the stats are in, and they are bleak. Alcohol use is the seventh leading risk factor for disease worldwide according to a massive study bringing together global datasets on drinking and disease. The study, which was published in 2018 by the Global Burden of Disease Collaborator Network and funded by the Bill and Melinda Gates Foundation, claims 'our results show that the safest level of drinking is none … This level is in conflict with most health guidelines, which espouse health benefits associated with consuming up to two drinks per day'.

Crucially, alcohol was the leading risk factor for the deaths of people aged between 15 to 49 years. It was not the direct cause of most of these deaths – tuberculosis, road accidents and self-harm were the three leading causes of alcohol-related deaths for young people. But according to the study's data, drinking made these deaths more likely. This is weird, evolutionarily speaking. The ages 15 to 49 are a crucial window for natural selection – things that kill you before you are old enough to have and raise kids are the most likely to be selected against.

Clites is an evolutionary thinker, so he was frustrated to discover addiction researchers with an evolutionary approach were rare. Exciting work is being done, however, mainly by evolutionary biologists outside the field of addiction medicine. In 2023 Clites was the lead author on a paper synthesising some of this work titled 'The promise of an evolutionary perspective of alcohol consumption'.[9] The paper reads as a plea to the research community to start taking evolution seriously when thinking about addiction, and it motivated me to get in touch with Clites. To understand his proposals, let's first look at what alcohol is and how species, including humans, have evolved in their relationships with it.

The rise of the drunken monkey

Yeast, much like love, is all around. It's on our fingers, our toes, it blows in the wind, and as such it lands on flowers and fruits. Fruit is a plant's way of enticing animals to spread its seeds. When fruits ripen, the sugar content increases, enticing hungry creatures to pick and eat the fruit, and hopefully drop the seeds somewhere they can flourish. Ripening can take some time, and as fruits tend to be imbued with yeasts, they have evolved anti-fungal defences, allowing them to mature before the yeasts devour their sugars. As they ripen, their defences weaken and yeasts begin to feast on the sugar-packed flesh. This fungus feast, called fermentation, creates waste in the form of carbon dioxide and ethanol.

Being the inventors we are, humans have found ways to control the fermentation process. But ethanol is the same substance whether found in beer, wine, spirits or a mixed drink. Even a prestige bottle of wine costing a collector hundreds of dollars gets its magic from glorified fungus waste.

Up until now I have been using the word 'alcohol' for what we usually call booze. But alcohol is a catch-all term for several chemical compounds, including compounds that are incredibly dangerous to consume, such as methanol. The only alcohol we can hack is ethanol, and that is the one found in fruits. Through this chapter I will use the words interchangeably.

Animals that have a fruit- or nectar-rich diet – frugivores and nectivores – regularly get exposed to ethanol. Nathaniel Dominy of Dartmouth College studies how wild primates find food and what they eat. He tested the alcohol content of various fruits that were not visibly fermenting in the forests of South-East Asia, and although their ethanol content was low, he tells me, 'I can't find a fruit with zero alcohol. It's omnipresent.'[10] In other parts of the world the natural ethanol content of fruits can be quite high: 'Sometimes it's getting to 4 or 5 per cent, like a normal beer.'[11] Of course a big pile of fruit on the ground might be more likely to ferment, but even ripe fruit can have an ethanol content of 2 per cent. Alcohol is often present in nectar and sap at quite high levels.

Fruits that have high levels of sugar tend to have a higher ethanol content. So, to get the calories represented by sugar, frugivores must cope with ingesting ethanol. As a bonus, the ethanol itself is calorie dense, containing around double the calories of the same mass quantity of sugar, although there is some debate about our capacity to unlock those extra calories.

Many primates, including many of our ancestors, are frugivores, and eating fruit leads to at least some exposure to ethanol. In the late 1990s this got Robert Dudley of the University of California, Berkley, wondering if humans' penchant for drinking can be traced back to our ancestors' taste for fruit, especially fruit that has started to get a little funky. Assessing the scientific explanations for addiction to alcohol in humans, Dudley noticed they tended to take the short view, assuming alcohol consumption was a relatively new advent in our evolutionary history.[12] Because it was a 'novel substance', ethanol was able to hijack our brain, and we had not yet evolved defences against it. But we come from a long line of fruit eaters who are regularly exposed to ethanol, so for Dudley, the idea that our lineage only recently met the drug did not seem likely.

In 2000 he published his first paper explicitly arguing that our taste for alcohol is directly connected to the survival benefit alcohol provided to our ancestors. How? Many primates live in tropical rainforests. These environments look lush but can be tough places to find a meal. Alcohol is smelly, you will know this if you have ever spilt a drink on yourself or walked into a room the morning after a party before the bottles and glasses have been cleared away. Did primates use the smell to track down fruit in the forest?

This idea became known as the 'drunken monkey hypothesis'. Dudley argued that his idea could explain our enthusiasm for alcohol. This is where context matters. A fart joke might make a room of kids laugh, but if you tell the same joke in a prudish workplace, you could get sent to HR. In the same way, traits that help a species survive in one environment can be highly dangerous in a new one. This idea is called evolutionary mismatch. Dudley argued that the same behaviours that highly motivated our primate ancestors to find a meal can lead to alcohol use disorder in a modern context.

It was an intoxicating idea and at first glance makes a lot of sense. But initially the primatology community was not convinced. Early critiques were laid out in a 2004 paper by Katharine Milton, also of the University of California, Berkeley, delightfully titled 'Ferment in the family tree: Does a frugivorous dietary heritage influence contemporary patterns of human ethanol use?'[13] The researcher surveyed her colleagues from around the world about the eating habits of primates. She was told primates tended to prefer fruits that were ripe, but not visibly fermenting. At the time she concluded there was not enough data to support the drunken monkey hypothesis, and that it was likely primates would be deterred by foods with a high ethanol content. However, since then studies of primate diets have become more detailed, so despite its initial stumble, the drunken monkey has picked itself up off the forest floor.

Wild primates are rarely found chowing down on big piles of fermenting fruit, preferring fruit that is ripe but not too far gone. So even if they do enjoy alcohol, there might be things about overripe fruit that turn them off – and who can blame them. I like a cold beer or a gin and tonic but not half-fermented wormy fruit. But all primates have 'fall back foods', things they will eat when food is scarce but disdain at other times. So maybe being able to eat overripe fruit could help in a crisis.

In 2021, a study came out showing that fermented fruits were found to make up between 0.01 per cent and 3 per cent of the diet for a range of primates from Asia, Africa and South America. The fruits were a small part of the diet, so easily missed in those earlier surveys, but the researchers concluded that the consumption of these fruits likely expanded their diet, especially when food was scarce. They write: 'It is possible that the human propensity for fermented food consumption is rooted in this ancestral primate strategy.'[14]

The fall of the planet of the apes

So, there is alcohol in the environment, and primates do consume it in hard times. But the fact some species turn to booze as a last resort is not enough to explain our modern penchant for drinking. Enter the African apes.

Earth, or at least Africa, Asia, and Europe, truly once was the planet of the apes. Their heyday was in the Miocene Epoch, around 23 million to 5 million years ago. This fascinating chapter in Earth's history was well after the meteor that took out the non-avian dinosaurs and before the ice ages we associate with woolly mammoths, Australian megafauna and the Neanderthals. In the Miocene the first dog, bear and horse-like creatures appeared on land and the ancestors of kangaroos began to hop.[15] There were also over 100 species of apes, including the small long-armed *Oreopithecus* from Italy and the larger *Danuvius guggenmosi* from Germany, who could likely walk upright.[16]

This Eden of the apes could not last forever. As the Indian subcontinent smashed into Asia and the Himalayan mountains formed, the climate became drier. The vast forests preferred by most apes started to contract, and grassy savannahs emerged, putting the apes under stress. In a harsh new world, only a few ape species made it through. In the jungles of Asia, the ancestors of orangutans and gibbons survived. And so did the ancestor we share with gorillas, chimpanzees and bonobos in Africa.

What gave our ancient African ancestors the edge? These were tough animals that had employed a generalist lifestyle to make it through millions of years of tumultuous environmental change on the continent.[17] Around ten million years ago, our shared ancestor underwent a random genetic mutation that could have been one of the key factors in its survival. One change to the ADH4 gene meant this new allele now coded for an enzyme that allowed for alcohol to be metabolised 40 times more efficiently.

The timing of this mutation is key. Ten million years ago was also about the time our shared relative is thought to have swapped an exclusively tree-dwelling lifestyle for a more grounded existence. While chimpanzees, bonobos and smaller gorillas often climb and sleep in trees, they spend much of their day on the ground – where fallen fruit is more likely to be fermenting. The ADH4 finding was published by Matthew Carrigan of Santa Fe College and colleagues in 2014.[18] The existence of this mutation narrows the focus of the drunken monkey hypothesis. While frugivorous primates all have some exposure to low amounts of alcohol, the mutation allowed our ancestors to unlock a new food source in the jungle. They could consume quantities of fermenting fruit on the forest floor that would have made our monkey cousins wasted without getting drunk.

Sometimes a mutation spreads because it gives individuals a reproductive advantage. But evolution can be wishy-washy and mutations can just get lucky and spread due to random chance. Finding a gene that helps us metabolise ethanol does not necessarily imply that our ancestors employed it, and in so doing evolved other traits that attracted us to booze.

So, what suggests alcohol appreciation was selected for? By being able to digest high-alcohol foods, apes might have been able to take advantage of a resource less accessible to their competitors, such as monkeys. This is especially important as monkeys can eat unripe fruit as a fallback, but apes are not able to digest it well, often favouring shoots and bark instead. Apes that were rewarded by their brain chemicals for indulging in fermenting forest fruits might have been more motivated to find them, eat more and survive, possibly selecting for the increased reward circuitries that alcohol triggers.

The next hurdle the drunken monkey had to stagger over was that of animal preference. Was there evidence that apes would consume higher ethanol meals intentionally or, as was suggested in 'Ferment in the family tree', would high ethanol content dissuade consumption? A community of binge-drinking chimps put this idea to rest.

A 2015 study authored by Kimberley Hockings of the University of Exeter and colleagues named 'Tools to tipple: Ethanol ingestion by wild chimpanzees using leaf-sponges' presented 17 years of data recording the drinking habits of chimpanzees in Bossou, Guinea. In Bossou, humans set up containers to capture the tapped sap of raffia palms. The sap ferments and has been recorded reaching between 3.1 per cent and 6.9 per cent alcohol, so in some cases it was stronger than a regular beer.

Cheeky chimps used leaves as tools to extract and then consume the fermented sap. Drinking events were recorded 51 times over the 17 years. The authors note that on occasion some individuals 'displayed behavioural signs of inebriation' and 'some drinkers rested directly after imbibing fermented sap'. The sap containers were not a constant presence in the chimps' home range, so were not a reliable part of their diet, nevertheless, the scientists wrote:

> We observed individuals repeatedly consuming fermented palm sap – often in large quantities – suggesting that accidental ethanol ingestion is unlikely. Our results clearly indicate that ethanol is not an absolute deterrent to chimpanzee feeding in this community.

Apes are not the only animals who fancy a drink or have the enzymes to deal with it. Lemurs are primates that veered off from the rest of the family tree before the branch that led to monkeys and apes. They only live on the large remote island of Madagascar, where there are around 100 different species, many with specialist adaptations that allow them to take advantage of specific resources. Most lemurs are unbearably cute, but not so the aye-aye. These primates look like balding brush-tailed possums with bright yellow eyes. But their most disturbing feature is their bony extra-long black middle finger, specialised for hooking grubs out of tree branches.

Most of the year, aye-ayes are grub-eating specialists, but for a few months of the year they eat the nectar that froths and ferments in the large flowers of a Madagascan palm called the traveller's tree. Why aren't they falling drunkenly out of these trees? The aye-aye share the same genetic mutation allowing for alcohol metabolism that we do.[19] This implies an ancestor of the species independently experienced the same mutation that we inherited. Meaning the same mutation has occurred twice in the primate family tree. They use their fancy enzymes to drink fermented nectar when it is in season. But is it just the sugar they like? Or the booze?

To test whether aye-ayes liked a tipple, Dominy set up an alcohol-tasting experiment with captive aye-ayes. Holes were drilled into a table, and alcohol of varying percentages was poured in. He explains that 'the animals would systematically go around the table periphery and smell each hole. And then once they found the highest [ethanol] concentration' they would dip their creepy long finger into the hole and lick it.

'They would empty the cups of alcohol very quickly,' Dominy says.

His interpretation? They loved it. From reviewing the

literature and working with aye-ayes, Dominy concludes that primates, including humans, have a strong preference for alcohol because consuming it has provided us with benefits. However, he does concede that humans 'overdo it'.

Alcohol tolerance has evolved in many species not related to us. Bats, tree shrews, fruit flies, nematode worms and many other animals also have fruit- or nectar-heavy diets. All these animals have been found to have adaptations that help them metabolise alcohol. Bats tend to fall into one of two categories, insect eaters or fruit eaters, aside from a few outliers like the blood-loving vampire bat. Frugivorous bats often have mutations to the same gene as the African apes and aye-ayes. Exactly how this gene varies is different among species – as is their alcohol tolerance – whereas insectivorous bats tend not to have the mutation at all. Tree shrews have incredible alcohol tolerance, consuming heroic amounts of ethanol-rich nectar and appearing to stay sober. However, their genetic pathway to metabolising it seems different to ours.

Not all fruit lovers can hack booze. Elephants might be heavyweights in their environments, but they are lightweights when it comes to alcohol. African elephants will often gorge on large amounts of fruit fermenting on the ground. Stories of them getting wasted and going on rampages are the stuff of legend. In 2006 a paper suggested these stories were likely myths. Extrapolating from human physiology, the authors argued that African elephants are simply too big to get drunk on fermenting fruit.[20]

But we know it's not the size of the animal that counts, it's what they have in their genes. Genetic research published in 2020 shows elephants lack adaptations to metabolise ethanol, meaning their threshold is far lower.[21] Why do they consume so much fermenting fruit if they do not have the metabolism to deal with it? It's possible that it is a lot safer for a large animal like an elephant to be drunk than a vulnerable small bat that must navigate the night sky. So if an elephant was born with a spontaneous mutation that limited ethanol's intoxicating power, that mutation would not be as likely to be selected for as it would be in a bat. This interspecies comparison shows that a species' relationship to alcohol is influenced by many things, including diet, risk of predation, environment, and even whether they fly, climb, or walk.

Species that have evolved traits that allow them to consume ethanol tend to have distinct changes in their digestive system. In humans booze starts getting metabolised as soon as it hits our throats, and this process continues in specific parts of the system. The pattern is similar across many frugivores. But it is not in animals like rats, who do not share a long evolutionary love story with alcohol.

However, it is rats and mice with a teetotal past that are used in many studies of drinking. The research often involves a protracted period of 'addicting' the animals to alcohol before then studying how addiction changes the brain, body, or behaviour. Clites tells me, 'Rats and mice can be great for studying addiction once you're already addicted', but they are not great if you want to understand who is vulnerable to alcohol use disorder after exposure. Clites believes studying species who evolved with ethanol in their diets could provide new insights into who is predisposed to alcohol use disorder and why.

Drinking in humans

I can see how enjoying alcohol as a foraging primate in a tropical forest ten million years ago might give a species an edge. But later early human ancestors such as *Homo ergaster*, who emerged about two million years ago, were walking the savannah where sugary fruits were harder to come by. Modern humans, *H. sapiens*, have been around for at least 200 000 years and evolved in similar grassy environments. Currently, there is no hard evidence to show that *H. ergaster* or our very early *H. sapiens* ancestors brought a culture of drinking with them onto the savannah.

Traces of what might be the world's first beer was found at a 13 000-year-old burial site of an ancient foraging semi-sedentary people called the Natufian. The site is in a cave in modern Israel.[22] Fascinatingly, these beers were made with wheat several millennia before the grain was domesticated. Between 9000 and 8700 years ago, drinking was part of the culture in what is now Qiaotou, China.[23] Chemical analyses of pottery shards found at a burial mound showed that the broken vessels had held beer. This research hints at ritual drinking. There is something moving in knowing that these ancient peoples were drinking to honour their dead, as many of us still do today. Alcohol is a major part of most human cultures, so much of the archaeology around it centres on finding the origins of brewing and early evidence of the practice.

This timeline implies a dry spell of several million years – but this seems too long between drinks. The argument that humans began drinking around the time they started farming is made by British palaeontologist and evolutionary biologist Nick Longrich. In an article for *The Conversation*, he writes: 'Archaeology suggests alcohol and drugs date back millennia, to early agricultural societies. But there's little evidence early hunter-gatherers used them.'[24] This view does not exclude early alcohol use, but it does minimise its role in the human story. I worry it overlooks alcohol use in cultures that did not share the agricultural traditions of parts of Eurasia, Africa and South America over the past 11 000 years.

People have lived on the Australian continent for at least 60 000 years, with some estimates almost doubling that timeline. While hunting and gathering does not adequately describe the level of geo-engineering and land management that was developed here, it is the way First Nations life has often been framed by anthropologists. Around the continent people innovated diverse and specific fermentation technologies.

Where I live, in lutruwita/Tasmania, Aboriginal people made a drink called *way-a-linah* by tapping a eucalypt species called the cider gum and allowing sugary sap to ferment in bowls carved at the tree's base. In other parts of Australia alcohol was fermented from banksia flowers brewed in large bark-lined troughs. Records detail that this sweet mead was enjoyed during feasts. In what is now the Northern Territory, *kambuda* was made from fermented pandanus nuts.[25] And this list of drinks is just a taster.

Evidence of diverse, specific, and ingenious fermentation technologies in pre-invasion Australia is not direct evidence for drinking throughout the more than 200 000 years of human history. But it does hint to me that even without the kind of agriculture that started popping up around 11 000 years ago, people seem to find a way to drink. Without pottery shards or written records, we will never know exactly how and if alcohol was brewed by most non-agricultural cultures through time, but – and this is just my hunch – I suspect we have been getting drunk for fun, feasts, and serious ceremony for much of our species' history.

What is indisputable is that alcohol and agriculture seem linked through time. This connection is so strong that there is an ongoing debate as to whether early agriculturists were motivated to take up farming to provide themselves with more alcohol, rather than more food. That debate is beyond the scope of this book. But the fact remains, from the time of the first farms, a lot of energy was going into alcohol production. So, how has evolution operated on that tendency?

The good side of drinking

Humans are deeply social. Without people around us, we quickly fall apart. Social isolation and loneliness are bad for our health, decreasing cardiovascular and mental health and increasing chances of mortality by any cause. Social connections are important, allowing ideas to spread and increasing social safety nets. So there are both group and individual benefits of strong social connections, suggesting that natural selection might act on things that create group cohesion.

Connecting with another person is a tricky business. Famed biological anthropologist and evolutionary psychologist Robin Dunbar has spent his career focussing on social relationships, including how they are formed and maintained. He explains that making friends is a neurochemical experience, and a key group of chemicals our body makes to facilitate connection are endorphins.[26] Ah, here we are – back to the endorphins. Their release not only makes us feel great but helps us trust and bond with others. Dunbar and colleagues have found that our brain is triggered to release them by wanted physical touch like stroking or cuddling, telling stories, dance and laughter. But two things that really unleash a mega-load of endorphins are group singing and drinking.

The role of substances in group formation had been relatively overlooked until some of Dunbar's collaborators mentioned to him that alcohol kickstarted the same endorphin system he had spent years studying. He tells me that 'in social contexts, across human cultures as a whole, with some exceptions, alcohol forms a major component of feasting'. This made him suspect that alcohol had been maintained in our cultures because of its social bonding effects.

Dunbar contends that alcohol would have played a role in hunter-gatherer societies through time, although its use would have been generally limited to 'once in a blue moon' events and get-togethers of extended networks. He says, 'It ramps up this sense of community bonding, which is hard to do', and suggests that communities and individuals that had deeper social bonds might have had a survival edge.

Without settled agriculture, Dunbar believes alcohol use, while important, would have been limited because of the logistical challenges of large-scale brewing. The advent of large vessels and cultivated crops allowed for high volume production, really getting the party started. Living together cheek by jowl, or field by field, is stressful. So as villages started to form, Dunbar believes alcohol's social role became more crucial, musing:

> You need mechanisms to stop people murdering each other ... small scale societies have evolved various kinds of social institutions to deal with that, one of which is feasting ... I suspect that alcohol became very much more important in that particular context, simply because it's very fast in terms of the bonding process.

This is a nice story, but understanding the brain chemistry of alcohol does not prove drinking helps people form and maintain relationships. And it points to the complexity of this story, as we know alcohol can also increase violence. Humans, like all animals, exist in a habitat, so to understand how alcohol interacts with human relationships 'in the wild', people need to be studied where they are drinking. Dunbar decided to take his hypothesis out of the lab and into the bars and pubs of England. He wanted to know if social drinking impacted people's social networks. And whether the effect was the same in old-style community pubs, 'locals' that are quieter, have more regulars and lower alcohol consumption, compared with louder, drunker, inner-city bars that have fewer regulars. The study participants included adults of a diversity of ages, unlike similar studies that focused on students.

Dunbar's human drinking field research found 'that social drinkers have more friends on whom they can depend for emotional and other support, and feel more engaged with, and trusting of, their local community [compared to non-drinkers]'. This is a big deal, considering how vital a social network is to our survival. The result was arrived at by a combination of survey data, in-pub experiments and questionnaires, as well as data collected by observing people converse and socialise – imagine trying to have a chat in the pub while an anthropologist notes down who you talk to, how long for and if you touched your phone!

Dunbar's anthropological findings also showed that people who regularly drank at their local felt more connected to their community and had larger social networks than bar drinkers, indicating that context mediates the way alcohol may interact with social connections.[27] It is important to note that people who have larger social networks might drink more socially because of those networks, so causation is tricky to prove. I wondered if living in the type of community that has a 'local' might be more conducive to long-term friendships. When I ask Dunbar about this, he tells me, 'The point is precisely that community pubs are social hubs for the village and/or community.'

The endorphin system helps us trust people. Dunbar's fascinating in-venue experiments, such as getting punters to rate the trustworthiness of strangers' faces, did not show that people engaged in moderate social drinking were more trusting of strangers. This led the experimenters to speculate that 'alcohol is more likely associated with the maintenance of existing relationships than with the initiation of new ones with strangers'. I suspect this might change when people are smashed, however.

While a drink may not make us more trusting of strangers, liquid courage might make us more willing to approach them. Chatting to Dunbar, he tells me the reduced inhibitions resulting from alcohol may help people make new connections, especially romantic ones, as drinking reduces the 'inherent social nervousness that we are all saddled with'.

But social drinking isn't all rosy. Dunbar identifies an inbuilt conflict involved in drinking, saying, 'Lowering your social inhibitions and just going for it opens you up to the risk you'll have trouble.' For some cultures the trouble is not worth the possible benefits. For example, Islam prohibits drinking, and many schools of Buddhist, Hindu and Christian thought are also anti-alcohol. While it is important to note that people may flout these rules, drinking is often lower in these cultures and is illegal in some countries. In these cases, how are connected communities formed? Dunbar describes to me two methods, which I think of as falling into either replacement or repression strategies.

Group drinking is not the only way to bind communities. Dunbar explains: 'Creating these large-scale bonded societies is very, very challenging and there isn't a simple answer to how to do it. That's why we have so many different behaviours involved in what I call the social tool kit.' These can include taking other substances, such as coffee, which Dunbar says 'almost certainly' triggers endorphins and is widely consumed, even in many dry cultures.

Ingesting something is not necessary, however. Drinking and other drug use among Icelandic teens was slashed when policymakers decided to find healthier ways to spark their brains' endorphin systems. Parents were encouraged get their kids involved in sport and other activities that triggered endorphins and were given vouchers by the government to subsidise their involvement. They were also given explicit instructions to nourish their relationship with their kids and make time for their social bond. Teens were also put under curfew. The replacement and connection strategy was so successful at reducing drug and alcohol use that it has been picked up by various cities around Europe, where it has found similar success, even without the curfew.[28]

For those with a spiritual bent, 'modest amounts of religious ritual and religious attendance' is a great way of bonding a community, Dunbar says. It makes you 'feel more contented and happier with your lot in life'. However, he warns that massive doses of religion can also be unhealthy for individuals and communities.

Dance and group singing forge bonds. Whether you're headbanging at a Christian rock concert, engaging in the Sunni tradition of Qawwali singing, or dancing the whirling dervish, your brain will be releasing chemicals that help you connect. Dunbar explains that these 'activities whip you up into a state of trance and a sort of magical mind state'. Just like drinking, 'the problem is that it can very easily spill over if it's not very tightly controlled into all sorts of dodgy social behaviours.'

Enter suppression. Some cultures take a dim view of most things that spark these social bonding chemicals. Dunbar points to the very strict Scottish Free Presbyterian Church, which disapproves of dancing (even non-touching line dancing), does not allow musical instruments in church, and generally encourages strict social control.[29] While Free Presbyterians do engage in some endorphin-sparking behaviour, such as group singing of psalms, Dunbar suspects the active suppression of these endorphin-promoting behaviours is why this group has remained small and limited to contexts where social pressure can force people to 'toe the line'.

It can be tempting to think that evolution selects for an optimum system, but it can sometimes be a lot more lackadaisical. Selection results in 'good enough' systems, especially when there are competing selection pressures. Many of the things we do to ensure our social bonds are strong – which is essential for our survival – have downsides. Take dancing. In the harsh environments some of our ancestors lived in, dancing represents the burning of a lot of hard-won calories. Sure, it would be great if we evolved a way to trigger the endorphin system and enjoy the other social benefits of drinking without the commensurate risks. But if the benefits outweighed the risks – resulting in more people successfully raising kids and their kids doing the same – well, booze, dance or prayer might not be perfect, but evolution will take it.

When drinking is selected against

Dunbar's work shows how a taste for alcohol may be preserved, but can the scale ever be tipped so the costs of alcohol result in the selection of traits that deter drinking? Ben Clites says this has happened more than once in the agricultural world. For some people, drinking even small amounts brings on nausea, heart palpitations and a red face. This is often called the 'Asian flush syndrome', so named because it is most common among people from south-east China, Korea, and Japan, though it is also present at high rates among Ashkenazi Jews.

Alcohol metabolism is a two-step process. It is first converted into acetaldehyde, which is incredibly toxic. Then this is converted to the safe acetate, which we pee out. People who inherit one version of the gene associated with flush struggle to do the second step, and people who inherit the allele from both parents cannot do it at all. In both cases the extra toxic acetaldehyde builds up, making them feel sick. This syndrome might be unpleasant, and drinking is extra bad for the health of people with it, but it massively reduces the risk of developing alcohol use disorder.

Fascinatingly, the alcohol use disorder protective gene is found at highest concentrations in the parts of south-east China where large-scale fermentation of rice wine started 8000 to 12 000 years ago. Industrial alcohol production slowly moved west over thousands of years. When the spread of rice wine tech and the spread of the protective allele are plotted on a map, an intriguing pattern occurs. The earlier a community in south-east China started making rice wine, the higher the frequency of the protective allele.

Scientists have been able to show that the frequency of this allele is a result of selection and not random chance. This has been done by analysing the genome around the allele in affected populations, and research that involved exhuming and genetically testing ancient skeletons and teeth.[30] While it could also have been selected because it seems to reduce the risks of some cancers, Clites says the evidence is 'convincing' that the allele was selected because people in the region who got sick after drinking were doing better than the heavy drinkers.

A person's genetics can also increase their susceptibility to alcohol use disorder. This is also related to alcohol metabolism. Clites explains that 'the genes that are better at metabolising alcohol and reducing its effects predispose you towards becoming an alcoholic'. So, people who can drink their peers under the table are more likely to run into problems than lightweights. He is fascinated by the genetics of alcohol use as some members of his family have the 'Asian flush syndrome' gene and hate drinking, while others have a troubled relationship with the bottle that he suspects results from their DNA.

Addiction and alcohol abuse are also highly associated with trauma, cultural dislocation, and other environmental factors like the availability of alcohol. Clites warns against looking at a population with high levels of alcohol use disorder and assuming it is the result of genetics. He points to the example of Indigenous communities in North America, where there are high rates of alcohol use disorder. When the technology made it possible, some scientists thought they could 'mine' the communities for genetic factors that led to the disorder. However, the limited genetic studies done on these communities have shown that they do not have an increased genetic susceptibility to alcoholism compared to people of European ancestry.[31] This is not to say that there isn't a diversity of susceptibility within the communities, or that some communities may be more at risk genetically, but as Clites observes, 'there's a huge confound of obvious socio-cultural factors'.

The danger of alcohol is intensely culturally and situationally mediated. Russia experienced an extreme alcohol crisis after the fall of the Soviet Union. This followed loosened restrictions, cheap vodka and, I presume, the stress millions faced when they were plunged into poverty. One large study of three Russian cities found that 52 per cent of deaths for people aged between 15 and 54 in the 1990s were associated with alcohol use. Excess deaths were high, the authors write: 'At the death rates of the year 2000, the probability that a 15-year-old man would die before age 35 years was almost 10 per cent.' In the rest of Europe that figure would have been 2 per cent.[32] In circumstances like this, it is easy to see how genes that were protective against over-consumption of alcohol could be selected for.

In crises and after trauma, drinking rates often spike. Some cultures glorify binge drinking, whereas others encourage moderate social drinking, and places like Australia combine the two practices. Through time, low-alcohol beverages have been drunk throughout the day, reducing infection as ethanol can ward against bacteria in contaminated water. In some times and places, low-cost hard spirits have been available at most shops. I can imagine the evolutionary scale rewarding a thirst for alcohol and a wariness of it has tipped through time and place, sometimes selecting for the tendency to tipple, sometimes selecting against it.

Clites is hopeful that our culture's relationship to alcohol is evolving: 'It's been with us for thousands of years at this point, it's really ingrained in sacred parts of our culture, it's hard to get a clean divorce from.' But he does hope that a shift to seeing addiction as a disease will encourage more compassionate responses and treatments. His call for an evolutionary approach to alcohol research is one that acknowledges our long history with the substance – both its beauty and its pain – and uses the window of our past to guide our future.

CHAPTER 8
Why do we sleep?

Seemingly fast asleep in a PVC pipe in a tank the size of a single bed is – a shark.

This is Edna the epaulette shark. At 68 centimetres from tail to tip, she is one of the many study subjects living at the Rummer Lab at James Cook University in Townsville, a sweltering city on the Queensland coast next door to the Great Barrier Reef. Jodie Rummer and her PhD student Aaron Hasenei have let me into their humid research facility. The tanks are set up like bunks at a budget backpacker's, and each of their inhabitants is snoozing, despite it being 11 a.m. Edna is a favourite among the research team. She was their first shark, her bunk is by the door, and she has a reputation for being sassy.

Rummer and her team are studying epaulette sharks to understand how their bodies will adapt as ocean temperatures become more extreme in the reef's tropical ecosystem. A surprising direction of their research has been to focus on sleep, a behaviour that may prove critical for this species' survival in a warming world.

Sleep occurs across the animal kingdom, but it poses a paradox. If Edna was snoozing on the reef, she would be less aware of the risks around her. So why, in a world full of danger and predators, has a behaviour persisted that leaves individuals so vulnerable?

Let's look at what we do know about sleep – and start with sharks.

Sharks are ancient. There were sharks swimming in the ocean before there were trees on land. Humans shared our last ancestor with sharks around 450 million years ago. The sharks' lineage stayed in the ocean, developed a cartilaginous skeleton, and diversified into rays and an otherworldly group of fish called chimeras.[1] Our branch of the vertebrate family tree, which we do not share with sharks, includes bony fishes, amphibians, reptiles, dinosaurs, birds, mammals and, of course, humans. The shark lineage changed over time, with species rising and falling, but their basic body plan stayed stable. Rummer admires this about the creatures, exclaiming that they are 'just so robust … surviving five major mass extinctions'.

Sharks are about as different from humans as a vertebrate can get. If there are elements of sleep we have in common, they may have occurred in an ancient shared ancestor. The other possibility is that these traits are so important they have evolved at least twice on our separate branches of the family tree.

A few minutes into my tour of the lab, our chatting wakes Edna up. She starts swimming around her tank, popping her head out of the water, hoping for a snack. This hungry shark does not look very – well, 'sharky'. She is slender with a long, elegant tail, and is very flexible as she bends her body up to check us out. Epaulette sharks are only found on the Great Barrier Reef and are specialised for life in shallow coral habitats. The shark's dappled black spots on a light caramel body help her camouflage on the reef and give her an effortless beige chic worthy of any old-money fashion label.

Edna's underside is flat, to allow her to lie on the sand, and she has four round, pancake-like fins positioned where the legs on a lizard would be. Incredibly, she can use these fins like legs to walk out of the water, a crucial skill when navigating shallow tidal systems.

'The biggest epaulette shark I have ever seen was just under a metre long – that's our "Jaws",' Rummer tells me. Their size makes them vulnerable in the shallows – 'sea eagles swoop down and grab them' – so during the day they hide in the complex structure of the reef and snooze, venturing out at night to hunt unsuspecting sand worms.

The sharks' home, the Great Barrier Reef, stretches south from the northernmost tip of eastern Australia. It is made of more than 2500 individual reefs, and most of the World Heritage-listed area is protected by a marine park larger than New Zealand.[2] But the reef is under threat. Runoff from land clearing and agriculture is combining with rising temperatures and other pressures to endanger this unique ecosystem.

Rummer's aim is to understand how global heating is affecting the sharks. To successfully do this she had to work out how to care for them in the lab. One of the first things she realised was that their activity varied drastically depending on the time of day. They would be active at night, and just want to snuggle into their PVC pipes during the day. This predictable 24-hour pattern of activity is called a circadian rhythm or body clock. It is mediated by light and shows up in life forms as diverse as bacteria, trees, and of course these sharks. For humans the circadian rhythm is one of two processes telling us when to sleep, and it is the reason we get jetlag. Patterns of daylight and dark influence when hormones such as cortisol and melatonin are released, influencing our feelings of wakefulness and sleepiness. Our body also senses how long we have been awake. When it's too long since our last snooze, we feel tired.

This mechanism is why, if you were to stay up all night, you might ignore your circadian rhythm and sleep during the day. This is called homeostatic rebound.

Attempts to change the circadian rhythm of the sharks were unsuccessful. If the researchers were going to mimic the conditions the sharks lived in, they would have to let them rest during the day. Suddenly the question 'Do sharks sleep?' became incredibly important to the design of their experiments.

Wait a second – do sharks sleep?

For a long time it was believed that sharks never slept. This added to their fearsome reputation as ancient, amoral, sharp-toothed predators who did not rest and could strike at any moment. But in recent years this consensus has started to crumble. To understand why sharks sleep, we first need to understand what sleep is.

For Rummer, sleep means having the aircon on to stay cool, cuddling with her partner in nice sheets and resting.

'I've got this nest that I make, and it's my recovery, it's my relaxation and that's essentially what any animal is doing when it's sleeping,' she tells me.

If a fellow scientist was to hook her up to a machine that measures the electrical activity of her brain, an electro-encephalogram (EEG), while she was in her nest, they would find that as she fell asleep her brain activity changed. There are multiple stages in human sleep, but they are broadly categorised as rapid eye movement sleep (REM) and non-rapid eye movement sleep (nREM). REM sleep is when most of our dreams occur.

Complicated and cumbersome EEGs cannot be used with most animal subjects. So, researchers have expanded the way sleep can be described to include other behavioural criteria. When we sleep, we do not respond to things that would usually get our attention. This is why fire alarms are loud. A sleeping animal will not react to things that would usually pique its interest when awake. This is also true for those hibernating or in a coma, but it takes a lot more to rouse them from those states than it does to wake someone up.

Rummer describes her bed as her 'nest' and prefers sleeping there. This too is common among animals, who often have site fidelity to their preferred napping locations. Most of us have one or more go-to sleeping positions. This is the same for many species, who stop moving and maintain a distinct posture when they doze off.

When animals seek extra rest after being deprived of it, this is a sign of homeostatic rebound. Just like us after a big week or a sleepless night, they are trying to catch up on lost sleep.

By using these criteria to spot sleep, it is easy to identify it across animal life, from tiny nematode worms and fruit flies to elephants. In the open ocean – or even the aquarium – these clues to spot sleep fall apart.

We breathe by moving air in and out of our lungs. This is how the carbon dioxide in our blood is exchanged for oxygen. Freshly oxygenated blood is then pumped around our body for our cells to use. In sharks and rays this exchange happens in the gills. Many species need to swim to breathe, as their movement pushes oxygen-rich water over those gills – just like when we breathe in and out. This means a white shark or a mako will never cuddle up in a comfy spot and drift off. But it does not mean they do not sleep. Rummer explains that for humans, swimming needs considerable focus, but for these sharks, swimming is just as effortless as breathing is for us. Their movement does not mean they are awake.

We know it is possible to swim and sleep simultaneously because dolphins do it. One hemisphere of their brain can sleep at a time, allowing them to keep moving and stay alert.[3] Hasenei, the PhD student in Rummer's lab, excitedly shares the tantalising possibility that great white sharks (*Carcharodon carcharias*) may use ocean currents as a bed. In 2016 a pregnant female made headlines when a camera on an underwater drone observed her navigating to a deep ocean current and swimming against it.[4] This would be a terrible idea if she was trying to travel efficiently, but it is a very low energy way to keep water flowing against her gills. Her mouth was wide open and Hasenei describes her as being in a 'trance-like state'. But without more research we cannot know for sure if this was true sleep.

Some sharks, like Edna the epaulette, can take a load off and stop swimming. She breathes by mechanically forcing water through her mouth and over her gills. Sharks with this skill are called buccal pumping sharks. Those who cannot stop swimming are ram ventilators. Rummer was keen to know if her sharks were sleeping when they were in their tubes or spending time in quiet wakefulness. As she was pondering shark sleep in her lab, a collaborative team working out of Australian and New Zealand universities were investigating just this question with draughtboard sharks (*Cephaloscyllium isabellum)* and Port Jackson sharks (*Heterodontus portusjacksoni*). These southern Australian and New Zealand sharks can also stop swimming.

The team tested whether the sharks were sleeping by measuring them against the behavioural criteria of sleep. In 2021 they published their findings: both species had times when they were less responsive to stimuli compared to other times. Just as we need our alarms to be loud, it took a lot to wake the sharks once they were sleeping.[5] In 2022 the team added to this research, showing that the animals' posture was different when they were not swimming but responsive to stimuli, compared to when they were not reactive.[6] While sleeping, they would lie flat on the sea floor in a 'recumbent position', but while just resting, they would be propped up on their pectoral fins. Although these species can sleep with their eyes closed – something the epaulettes do not seem to do – having their eyes closed was not a reliable indicator of whether they were awake or asleep.

Crucially, the researchers found the sharks were using less energy when sleeping than resting. Our bodies are constantly hosting a large number of chemical reactions that use energy – this is our metabolism. The sleeping sharks were found to have a lowered metabolism. As Rummer puts it, they 'stop being so expensive' for a while.

'Every species on the planet, except ourselves, is trying to save energy and be really resourceful,' Rummer explains. This might seem obvious – of course we sleep to save energy – but the 2022 shark research was the first to draw the link between sleep behaviour and the physiology of energy saving in our aquatic relatives. This research points to energy saving as one of the key functions of sleep across vertebrates.

We all need sleep to live

Sleep is essential to life. This has been illustrated by a series of experiments on rats that unsettle me so much they have made an appearance in my nightmares. After missing a night's sleep, we feel shocking. Our eyes are red-rimmed, we are distracted, emotional, and in this state getting sleep can feel like a life-or-death mission – which it just might be. Research undertaken by

Allan Rechtschaffen and colleagues in 1989 and again in 2002 showed that extreme sleep deprivation can be fatal.[7]

Rats were kept on a disk over water. When the rats nodded off, the disk started to tip, forcing them to stay awake. A control group of rats was kept on the same device but allowed periods of sleep. The physical changes in the sleep-deprived rats were extreme. They ate more, lost weight, broke out in lesions, and lost control of their body temperature. In the 1989 experiment the first rat died after eleven days and none survived past 32 days, although several were euthanised when 'death seemed imminent'.

This suggested that sleep was, for rats at least, akin to eating, drinking and breathing – impossible to live without. Oxygen is essential for the chemical reactions constantly occurring in our cells. Water provides a medium for these reactions to take place. And food is our fuel. But exactly *why* we need sleep is a mystery.

This question drives the research of Yu Hayashi and the Hayashi Sleep Lab at the University of Tokyo. Reflecting on the dramatic results of Rechtschaffen and colleagues, he tells me, 'No one really knows why these rats died, so I think it's a fascinating question to ask … In our lab our ultimate goal is really knowing the function of sleep.'

Hayashi and his team are not in the business of rat sleep deprivation. Their study subjects are a lot smaller and are allowed to get shut-eye. Well, they do not have eyes. Because they are nematode worms. Meet *Caenorhabditis elegans*, or *C. elegans* to its friends. This nematode worm is tiny, growing to just 1.5 millimetres long, with a simple nervous system and a small genome to match.[8] Our last common ancestor with sharks was 450 million years ago, but, in terms of the history of life on Earth, sharks are our close cousins – we are both vertebrates, after all. Our last common ancestor with *C. elegans* was even older. While pinning down exactly when this ancient relative lived is tricky, we likely diverged from each other soon after the appearance of the first animal with a head.[9]

Understanding why sleep is essential to life is clearly a scientific challenge. Researchers have been trying and failing to puzzle it out for years. Hayashi believes the answer is 'beyond human hypothesis or imagination'. For many, this view might be reason to swap research topics, but for Hayashi it is 'highly motivating'. He does not think he can come up with the answers through the power of imagination, so he no longer uses hypotheses to drive his work. Instead, he trusts in the random creative power of a process called forward genetics.

You might have experienced the results of forward genetics when picking a bouquet from a florist or selecting a delish-looking vegetable from the supermarket shelves. It is a common method of creating new strains of plants. Hayashi explains that gamma rays are directed at the plant of interest, causing mutations in its genetics. This random process occasionally leads to a higher yield, a more beautiful flower or another desired trait. In agriculture that new strain will be selected and bred from.

'In the case of biology,' Hayashi says, 'what you do is look into the DNA and see which gene is actually damaged in this new strain.'

Roundworms were randomly mutated by Hayashi and his team, then observed to see if their sleep changed. Via a shared screen, Hayashi shows me his set-up for monitoring the sleep of hundreds of worms at a time. Each worm is placed in a tiny well in a larger tray, then the whole thing is recorded with a video camera. In the speeded-up footage, even I can see which worms are awake and which are asleep – the wakeful worms wriggle, and the sleeping ones do not. Some worms were 'long sleepers', not moving for almost twice the time of the other worms. Long sleepers turned out to have a mutation to a gene called sel-11, suggesting this gene was connected to the worms' sleep in some way.

Could these sleepy mutated worms shed any light on sleep in humans? Mammals, including humans, also have the sel-11 gene, although in us it is called the HRD1 gene, but let's just call it sel-11 to keep things simple. This means it was likely around in our common ancestor and survived in both the lineage that led to roundworms and the lineage that grew a spine and led to humans and sharks. Keen to see if the gene had retained its link with sleep over the millennia, Hayashi's team mutated it in mice. They found that mice with the mutated gene would have massively increased non-REM sleep.[10] Clearly, the team were onto something – but how could this gene change sleep patterns in these very different species? Diving into the scientific literature, they found that the function of this gene had already been researched.

Each cell can be understood as a factory producing proteins that have essential jobs either in that cell or elsewhere in the body. After a protein is assembled, it must be folded into the correct shape. In every factory there will be errors and in the cellular factory this can mean misfolded proteins. The sel-11 gene codes for a protein that is akin to the factory's quality control officer, destroying misfolded proteins so they do not clog up the factory floor. It seems that when the gene was mutated, quality control went out the window. Hayashi and his team confirmed this by checking the cellular function of their mutated study subjects, finding that 'Both in mice and *C. elegans*, if you don't have [working] sel-11, then you have more misfolded proteins.'

Wonky sel-11 leads to a cluttered cellular factory floor, but what's this got to do with sleepiness? Hayashi tells me that when cells start to accumulate misfolded proteins, they send an 'SOS signal to the brain'. This signal prompts the brain to sleep. He believes his research suggests that 'Sleep is probably a period that is very important for removing these misfolded proteins from our body.'

Because sleep is essential for brain function, a lot of sleep research focuses on sleep and the brain. But in evolutionary history, sleep came before brains. 'Even jellyfish sleep and they don't have a brain,' Hayashi tells me. Clearing out misfolded proteins is important throughout the body, so Hayashi believes sleep may have originally evolved for reasons unrelated to the brain. He emphasises that this does not explain everything about the shared evolutionary roots of sleep. But, because this protein is shared between such distantly related species, they might have uncovered 'a universal factor of sleep pressure'.

Hayashi's insight could be important for human medicine. An accumulation of misfolded proteins is associated with a raft of human diseases, including diabetes, cancer, and Parkinson's disease. Within the cell, many proteins are made in a structure called the endoplasmic reticulum, or ER. When misfolded proteins accumulate there, it results in ER stress.

This is how it can play out in diabetes: a person loses the capacity to use a protein called insulin, which is involved in absorbing sugar into the body. Insulin is made by cells in the pancreas, and they respond by ramping up insulin production. Soon the ER in these cells becomes overworked. Like anyone who is overworked, it starts making mistakes – in this case, by misfolding proteins. Eventually the ER stress becomes so great the cell starts to lose functions, reducing that person's capacity to make insulin at all. Diabetes becomes harder to manage when someone is getting poor sleep, and chronic issues with sleep increase a person's risk of developing the disease.

To me, Hayashi's research opens up a new way of understanding sleep. This work may lead to a definition of sleep that focuses on the rhythms of metabolic processes, such as the cleaning up of misfolded proteins, rather than broad observations of behaviour. This makes me think about sleep studies of species like white sharks or dolphins, who never stop swimming. But Hayashi tells me it could also help us to study sleep in organisms that do not move and are completely different from us, such as plants and yeasts.

Back at the Rummer Lab in north Queensland, I am learning how these cellular processes during sleep may be one of the factors that allow the epaulette sharks to walk out of the water. This is an extreme action for any fish, as it massively reduces their capacity to breathe. Even in the water, this species must contend with limited oxygen. It forages in tidal pools, which can become oxygen depleted compared to the surrounding ocean due to the respiration of the animals that live within them. Low oxygen is dangerous for the body, causing a build-up of destructive molecules.

To survive these extreme conditions, the sharks slow down their metabolic rate and protect their brain function. But even with these tricks, their bodies still have to recover from the effects of low oxygen. Hasenei tells me that he believes epaulette sharks 'are probably some of the best sleepers around'. This is because they need long periods of sleep to metabolise the harmful molecules that build up as they forage.

Both the Rummer and Hayashi labs are uncovering ways sleep keeps our bodies healthy and functioning at the cellular level. These are fundamental reasons why we need sleep to live. But at the end of my conversation with Hayashi I am still left with a question. Why does this action, the clearing of metabolic waste, need to be accompanied by a loss of consciousness?

'That,' he replies, 'we don't know yet.'

Did sleep technology make us smart?

Sleep keeps our bodies going. But when I think of sleep I think of my mind. I am not an early riser. Left to my own devices I go to sleep late and rise late. But while I was training as a producer at my local radio station, my shifts would start at 6.30 a.m. – or worse, 4.45 a.m. I would taxi in the dark to the studio clasping a can of iced coffee like my life depended on it. As soon as I entered the building we were on. My colleagues were bright eyed and bushy tailed, pitching stories, following leads, and showing me how to use computer programs. When we got to the studio, they navigated the switchboard and multiple screens as if we were at NASA's mission control. My exhausted mind felt like it was moving through sludge. I would take notes and try to follow, all while despairing I would never get it.

I only started to understand the job when I shadowed afternoon and evening radio programs that allowed me to get more sleep beforehand. Soon I too was operating the switchboard, breaking news, helping talent shine, and navigating the complex relationship between radio host and producer. Being sleep deprived decreases our cognitive skills, makes it harder to learn, makes us more error prone and impacts our mood.

Our brains are busy when we sleep.[11] Animal models show that the places where the brain cells communicate, the synapses, increase in number during the day and are pruned back at night. This process may be important to help us keep learning new information over time.

Being a species that lives by our wits, it is likely sleep had a major role in our evolution. David Samson, Associate Professor of Evolutionary Anthropology at the University of Toronto, has been trying to decipher the twists and turns of the journey to human sleep. He argues that a series of technological innovations allowed humans to become 'freaks of high-quality sleep', allowing us to unleash the cognitive and social skills unique to our species.[12]

To understand how our unique sleeping behaviour made us human, we need to go back and understand how our early ancestors slept, and what changed for us. Primates started popping up in the fossil record around 55 million years ago. These tree-dwelling creatures probably looked more like possums than monkeys, and are thought to have been nocturnal, resting in nest-like constructions or tree hollows during the day.[13] They probably slept a lot, as modern primates that are active at night sleep up to 16 hours a day.[14]

As the millennia went by, many primates got larger. These chunky not-yet-monkeys had to abandon tree hollows and instead sleep exposed on branches, a dangerous practice. Even now tree-sleeping monkeys sometimes fall to their deaths. Around 25 million years ago a new group of primates, the apes, entered the picture. They were large, smart and did not have tails. Their growing size soon tipped the scales against branch sleeping. But snoozing on the forest floor left them vulnerable to predators. This problem was solved by the invention of the 'nest' or 'sleeping platform'. Ape nests are single-use beds constructed in the trees. While emphasising that these may have been innovated multiple times by multiple apes, Samson imagines the inventor as some 'sleepy, genius ape' living 14 to 18 million years ago.

This culture of bed-building still exists among chimpanzees, bonobos, orangutans and gorillas. If you travel through great ape territory, their nests are often easier to spot than the animals themselves, although they may be swaying tens of metres above you. The creation of these beds is simple and quick. The apes will pull branches towards themselves, snap them, then weave them together in an action that Samson likens to manoeuvring the flaps of a cardboard box. They add to the comfort of their nests by lining them with leaves. Nest-building technique is passed down from mother to offspring, with young apes starting to sleep in their own beds after they are weaned. Samson says it is possible this innovation allowed for a 'great leap forward in cognition'.

How would this have worked? A nest is comfy, safe and helps an animal control its temperature, allowing the occupant to let it all hang out and 'relax in an insouciant position sort of cupped by a tree bed', Samson explains. This luxury means apes can, and do, indulge in deeper and higher quality sleep than branch-sleeping monkeys. As deep, quality sleep is so important for thinking, memory and learning, this technological innovation may have set the stage for greater cognition to evolve.

Moving the slumber party to the ground

Our human ancestors were walking on the ground long before they were sleeping on it. Over 4 million years ago, a genus of hominin called *Australopithecus* was walking upright in the African savannah – even leaving fossil footprints. Their skeletons show they were still impressive climbers – 'They could arm-hang for days,' Samson says. Their capacity to easily scamper back into the trees suggests they were probably still weaving nests at night.

Sleep is thought to have migrated to the ground as early members of the genus *Homo* – our genus – emerged. Around 2 million years ago *Homo ergaster* appeared on the scene. Whether *H. ergaster* and *H. erectus* are the same species is a matter of debate that I will leave to others. According to Samson, this *H. ergaster* 'looked human from the neck down', but their brains were far smaller than ours. To sleep safely on the ground, they would need a new strategy, and Samson believes this strategy was social.

Some apes nest within cooee of each other but sleeping is still a solo affair. Early *H. ergaster* most likely rested much closer together. This could have allowed for some members of the group to act as sentinels while others slept, reducing the risk of sleeping on the ground.

Homo ergaster would have had many behavioural differences from us. But Samson explains that the 'paleoanthropological evidence suggests they were likely in camps that are pretty close in structure to forager camps that we see today'. The persistence of this social structure has given Samson the opportunity to test this hypothesis and see whether the watchfulness of peers can provide enough safety to sleep on the ground in the savannah.

The Hadza of Tanzania, who we met in our discussion about grandmothers, live in small communities of around 30 individuals, although groups can be larger. Their homeland is where ancient modern humans are thought to have evolved, so they hunt, forage and sleep in environments not too dissimilar to those of our ancestors. But it is important to remember the Hadza are a modern people, their culture is not static, and they have many traditions and technologies different from those of our forebears. Still, the environmental challenges the Hadza face are ancient.

Samson worked with Hadza communities to study sleep in conditions like those of our ancestors. A total of 33 healthy Hadza adults wore sleep-tracking devices called ActoGraphs as they went about their daily lives for 20 days, with 22 individuals wearing the devices on any given day. Even though people did most of their sleeping at night, over the entire 20-day period there were only 18 minutes when the ActoGraphs measured all participants being asleep at the same time. Breaking the 20 days down into minute-length 'epochs', these 18 minutes represent only 0.002 per cent of analysed epochs. During most epochs multiple community members were awake.[15]

The Hadza were not organising someone to always keep watch. So how did they manage it? When I started working in morning radio I struggled, but as soon as I switched to evening shows I flourished, whereas some of my colleagues prefer red morning shifts. While we all have a circadian rhythm, exactly how 24-hour cycles of light and dark impact our activity varies. Extreme early risers are colloquially called 'morning larks' and those who stay up late are 'night owls', with most people being somewhere between the two extremes. Where a person falls on this spectrum is referred to as a chronotype. It is partially informed by genetics and tends to be variable within populations.

While some people might have a lifelong tendency to be late or early risers, the greatest predictor of this trait is age. If you have ever tried to organise a multi-generational holiday, you might have some insight into this. Younger people tend to stay up late and wake late, but as we age, we tend to become morning larks. This could lead to disagreements if Grandma is up bright and early, ready to see the sights, but the teens of the family refuse to get out of bed. Tensions rise as the older adults point out the teens stayed up late talking: 'Why couldn't they just get to bed at a sensible time?'

Instead of each blaming the other for their frustrating sleep habits, Grandma and the teens might be able to unite in blaming biology. Samson found this generational mismatch held true for the Hadza. While it could be a source of conflict on a family holiday, the scientist believes this generational variability in chronotype could have far more utility in forager societies, shortening the window of time when all members of the group are asleep.

Samson describes the attitude to a night's sleep in western industrialised countries as a 'lay down and die' approach. Our understanding of a good night's sleep involves not waking up, and if we do wake, quickly returning to sleep. But this was not the case for the Hadza. After bedtime and before the morning, some people would be awake. This was explained by 'wake-bouts', essentially people just waking up for a small amount of time and then going back to sleep. People's 'wake-bouts' would happen at random times, often just for short periods, so even in the dead of night there would be alert individuals in the group. The unofficial sentinel system that allows the Hadza to sleep soundly probably evolved around the time our ancestors left the trees.

The Hadza also enjoy a suite of technologies that improve sleep quality, many of which also stretch back to before *Homo sapiens* evolved at least 200 000 years ago. This tech paired with social sleeping may have set the stage for humans to enjoy some of the most luxurious sleep the world had ever known.

Exactly when humans controlled the use of fire is debated. It is possible *H. ergaster* was wielding flames over a million years ago, but the evidence of humans using fire only becomes common around 400 000 years ago, and may have become habitual still later.[16] Lighting a fire is an incredible hack for a good night's sleep as it deters annoying insects and dangerous predators, plus it keeps people toasty warm. Huts and other building technologies would also have increased sleep comfort and safety. Samson pictures these social and technological factors coming together like a protective shell. The best thing about this shell is that it is mobile – when a community moves, it can set up structures and fires again. And there will always be someone awake and able to alert sleepers to danger. Thus allowing humans to carry sumptuous, safe sleep with them as they migrated into new environments.

Early *Homo sapiens* had complex social lives, lots of learning to do, and state of the art sleep tech. With these factors in play, it would seem a safe bet to assume humans sleep more than other apes. But, if you did bet on this, you would lose your money: 'We do something insanely bizarre,' Samson says. 'We are the shortest sleepers out of any primate you ever recorded.'

In industrialised settings, adult humans tend to need between seven and nine hours of sleep. Samson's work with the Hadza, and other research with small-scale societies, suggests these populations get by with even less sleep than that.

Chimpanzees and orangutans out-sleep us, getting about nine hours of shut-eye a night. By analysing our body, brain, and group size, along with our evolutionary relationships, Samson's models suggest we should be sleeping ten hours a night. But, he says, 'We're not, that's the hint something weird is going on. And evolution was actively pushing us one way or the other.'

At first glance, our short sleep time is a paradox. But Samson believes he has an explanation for it, arguing for what he calls the 'social sleep hypothesis'. Sleeping in that protective shell allowed our ancestors to reduce the quantity of their sleep by increasing its quality, allowing us to become the 'freaks of high-quality sleep' he first described to me. Reducing sleep time meant there were suddenly more hours in the day for learning, shoring up alliances, coordinating hunts and other 'really cool human unique behaviour', as Samson puts it.

What's dreaming good for?

Our sleep is not just of a higher quality than that of other great apes, it is also structured differently. Humans have a higher proportion of REM sleep to nREM sleep compared to our ape relatives.[17] While some dreaming can occur in nREM, our vivid narrative dreams occur during REM sleep. This shift suggests that there was an evolutionary pressure actively increasing REM sleep. Significantly, it also suggests that increased REM sleep was not being actively selected against. Samson explains that for REM to occur, specifically the part of REM called phasic REM – when you are 'about as dead to the world as you're ever going to be' – a safe environment is necessary. This may have been provided by Samson's sleep shell, allowing long-REM-sleepers to survive and to pass on their REM-heavy sleep style.

Exactly why sleep is split into REM and nREM sleep is another evolutionary mystery. REM gets its name from the way our eyes dart around during this sleep phase. Our brains, which cool down during nREM, warm back up, our heart rate rises, and EEGs reveal our brain activity to be similar to that of the wakeful brain. For most species, the body is paralysed during REM.

For a long time, it was thought that having sleep divided into these two states was restricted to Earth's furry and feathered creatures. If this were the case, the division between REM and nREM sleep would have evolved independently in each lineage that led to birds and mammals. But recent research suggests it may be more ancient. Lizards, fish and even some invertebrates may have states similar to REM and nREM.

That creatures other than birds and mammals also experience REM and nREM may have been obscured by our human bias to centre our own experience. REM was first described in humans and researchers have defined it to fit a mammalian model. This made it harder to spot in species with less complex brains who do not share other features common to birds and mammals.[18] If scaled and shelled creatures also experience two types of sleep, divided sleep may have been inherited by birds and mammals via a common ancestor.

I have always found REM sleep specially intriguing due to its connection with dreams. While we can dream in other phases, REM is where most of our dreams occur. Dreams are bizarre, fleeting, and hard to study. We can agree that humans dream, but our experience of dreams is subjective, and hard to remember. Despite this, researchers can track the rough contents of people's dreams via surveys and self-reports. By pairing up with people who are consistently aware they are dreaming and can control their dreams – lucid dreamers – scientists are even beginning to understand how specific brain activity relates to a person's dream world.[19]

Determining the function of dreams is tricky. REM sleep has an important role in memory consolidation, but determining if dreams are part of that process is hard to assess. One possibility is that dreams are 'spandrels' – that is, by-products of other processes (in this case consciousness) that serve no direct function.[20]

The very bizarreness of dreams might be a clue to their function. This is the view of neuroscientist Erik Hoel, who launched his hypothesis for a theory of dreams in a 2021 paper. He writes that the 'sparse and hallucinatory and fabulist [qualities of dreams] make it unlikely that strategies or abilities or preparations that originate in dreams work at all in the real world'.[21] Hoel has taken inspiration from the problems programmers have encountered in training sophisticated artificial intelligence programs called deep neural networks, which are inspired by the human brain and are trained on huge amounts of data. When given the same data or task repeatedly, they become stuck in a rut and unable to generalise information. This limits their flexibility, as they cannot take the knowledge gained from undertaking one task and apply it to another. To nudge these AIs out of their ruts, programmers introduce random variables. This injection of chaos makes the AI more able to apply old skills to new problems. Hoel argues dreams might have the same function, injecting chaos into the brain in a way that allows us to generalise our skills and knowledge.

Samson has a different view and sees dreams as a low-risk opportunity to simulate varied emotional, physical and social challenges. He suggests dreams may function as a 'virtual reality analogue, where you get to run through challenges'. If this is the case, human ancestors who had longer periods of REM sleep might have gained a survival advantage by being able to apply dreamed solutions to tricky problems.

Dreams also have a strong link with trauma. In 2020, as Australia went into lockdown to stop the spread of Covid-19, I started to be plagued by nightmares. When I asked my mates, many of them were also having vivid nightmares. This phenomenon was not confined to my circle – researchers in many countries investigated the relationship between lockdowns and nightmares and documented it.

Researchers studying lockdown nightmares experienced by Canadian students found that male and female students both had increased nightmares over the pandemic, but the effect was more dramatic for female students.[22] They suggested that this uptick in negative dreams could link to our evolutionary history. Pointing to the simulation hypothesis, they commented that these nightmares might give students the opportunity to 'creatively "play out" low-risk, hypothetical threat simulations' during a time of increased threat and anxiety.

Nightmares are horrific to experience, but may help us process frightening situations, allowing us to keep functioning and thus providing a survival benefit. Samson thinks of REM sleep as 'your night-time therapist'. In therapy a patient might examine a frightening event and try to reduce the fear associated with that event. Similarly, during REM, frightening experiences can be simulated with the fear factor turned down. It is thought this helps create 'fear extinction', allowing people to return to things that frightened them without experiencing terror.

As lockdown ended, so did my nightmares. But for people with post-traumatic stress disorder (PTSD), nightmares can become chronic, often with the same terrible situation being played out repeatedly in their dreams. In people with PTSD there are physical differences in the brain during REM sleep compared to those without the disorder. One of those differences seems to be that people with PTSD do not experience fear extinction during REM – so the fear factor on their nightmares is turned all the way up.

Not everyone who experiences a traumatic event will develop the disorder. Vulnerability to PTSD can be influenced by a range of things, from the severity of the event, genetics, and, interestingly, how well a person sleeps. In an experiment published in 2020, good sleepers and people with insomnia who often have abnormal REM sleep were taught to associate certain pictures on a screen with an electric shock.[23] The next day the good sleepers showed signs of fear extinction, but the results were the opposite for the people with insomnia.

Chronic insomnia can be devastating. But Samson believes being highly alert at night probably has deep evolutionary roots. 'Your ancestors got you here because of that,' he says. During periods of high threat, there is adaptive value in trading off a good night's sleep and the therapeutic effect of REM for a state of increased alertness. This is still true in the modern world. A person living with caring responsibilities, violence in the home, or who sleeps rough, might be kept alive, or able to keep kin alive, due to their night-time vigilance. This is backed up by research showing that the safer a neighbourhood is perceived to be, the better its denizens sleep.[24]

However, the stress many people encounter does not represent night-time danger. An overbearing manager or a university exam is not going to attack you as you doze. Samson hopes that understanding the evolutionary function of sleep disorders will help to treat them by helping patients reduce fear and anxiety. An evolutionary approach may also be helpful in the prevention or treatment of PTSD, as many researchers are exploring how REM sleep can be targeted to aid recovery.

'Turn off the light!' Sleep in the modern world

When I speak with Samson I am sure he is going to describe how technology such as artificial light, smartphones and social media is destroying our sleep. The scientist has been developing a database of information about sleep quality from around the world. Having seen countless headlines about a 'sleep deprivation epidemic', you could knock me down with a feather when he shares his findings with me: 'Western sleepers are sleeping the longest we likely ever have and the highest quality we likely ever have since the advent of our species, which is wholly counter to the narrative.' At time of writing this data has not been published, so I am excited to see if it stands up to scrutiny.

Something we do have is a lot of is anxiety around sleep. Samson's ActoGraphs showed the Hadza had shorter and more fragmented sleep than people in the industrial west. But when asked how they felt about their sleep, they replied it did not worry them. Samson says, 'They love their sleep ... There's none of this anxiety associated with the act.' He believes this is partly driven by the demands put on people in industrialised cultures, pointing to the pressure most of us have to be cognitively competent for an eight-hour workday, lamenting, 'No hunter gatherers ever had to endure this ... it's tough for us to imagine the amount of inherent flexibility we might have had in the past, and that a lot of these small-scale societies have.'

In industrialised countries, many of us have lost the flexibility to sleep according to our circadian rhythms. 'We fetishise sleep and we should be thinking about circadian rhythms,' Samson says. 'Circadian rhythms are the backbone of it all.'

In recent years a barrage of research has shown that misaligned circadian rhythms are associated with a swathe of diseases from diabetes and heart disease to cancer. Individuals can try to keep their body clocks healthy by spending some time outside at the start, middle and end of each day and avoiding bright lights at night (this is when phones and social media are a problem). But many of us are forced out of our circadian rhythms by shift work, as I was while working in radio. Understanding and managing the health risks of such work is a larger societal project.

Bright lights and industry also impact the sleep of the animals living in our towns and cities. In brightly lit cities, songbirds will start their chorus earlier than they do in less urbanised environments. This suggested to Thomas Raap and colleagues that these birds might be getting less sleep. This was put to the test in a 2015 study of the great tit (*Parus major*), a charming small bird found in Europe and Asia. The team measured sleep in tits whose nesting boxes were near artificial lights and those whose nests were not. They found that sleeping near light resulted in the tits rising earlier and sleeping less than their well-shaded counterparts. Laboratory animals exposed to bright lights at night show disrupted sleeping rhythms, hormonal changes – including increased stress hormones and immune system changes – suggesting artificial light might harm wild animals.[25]

Back at the shark lab, Rummer and her team believe that sleep will be a crucial tool for epaulette sharks if they are to survive in a warming world. Climate change is driving ocean heatwaves that bleach coral and put the animals of the Great Barrier Reef under immense physiological stress. Humans, along with birds and all other mammals, are able to keep our bodies at a constant temperature – what is often referred to as being 'warm blooded'. The body temperature of sharks, however, is controlled by their environment. The warmer they are, the faster their metabolism works – Rummer says it's 'just physics'. As the species uses sleep to deal with the by-products of their metabolism, one of the questions Rummer ponders is whether they are getting the sleep they need to complete these processes during the day.

Coral is essential to the reef, and thus the sharks' survival. The first recorded mass bleaching to hit the reef was in 1998. By early 2024, it had bleached another six times. While some corals can bounce back from a bleach, many do not make it, eventually crumbling away. The sharks' idea of a safe and comfortable bed is a complex coral matrix, and if reefs are threatened, so are their hiding places.

Rummer worries that coral loss may also impact the circadian rhythms of the sharks. It is thought that epaulette sharks lays their eggs deep within the dark matrix of the coral. In Edna's tank there is a fenced-off circle with lines of string crossing it like washing lines. Each line has one of Edna's smooth deep brown eggs pegged to it. 'Our hypothesis is that they are in complete darkness in nature,' Rummer says. The researchers are now experimenting to see whether exposing these eggs to different light conditions, as may happen if the reef continues to deteriorate, will impact the sleep and circadian rhythms of the hatchlings.

As the world around us rapidly changes, so will our sleep. Many of us can use technology to avoid the environmental changes wrought on the world that influence sleep. Beds, air-conditioning, and secure homes are likely boosting sleep quality for those who can afford them. But our work and caring demands, plus the temptations of technology, often insist we operate outside of our body's rhythms. It is becoming clear that the species we share the planet with, even unlikely ones like Edna the epaulette shark, also need a good night's sleep.

CHAPTER 9
Why do we have inner lives?

While swimming along at Hamelin Pool in Gutharragudu, I keep getting distracted by the fish. They are not abundant, but in such a sparse, hypersaline environment they stand out as flashes of light and speed. Silver shoals shelter under the bulbus tops of the stromatolites, tiny eyes seemingly following me as I glide past. A camouflaged football-sized fish scoots out from under me, finning away into the aqua beyond. It's an exquisite afternoon. The sky is blue and throwing shafts of sunlight down through the clear water. Compared to the ancient stone structures being painstakingly built by the microorganisms covering them, the fish seem lively, modern and miraculous.

It took a long time for this perfect scene in the pool to be possible. The universe roared into existence 13.8 billion years ago with a big bang. Simple small atoms like hydrogen formed. Over time stars, planets and galaxies winked and spun into being. More elements that make life possible, like carbon and zinc, were cooked in these stars or pressured into being as stars died or smashed together. Around 4.5 billion years ago, the little planet we call Earth formed. It was a scary place at first, bombarded by asteroids containing minerals and, crucially, enough water to create oceans. Eventually, things settled down a bit, and the stage was set for life to be built from the ingredients created in this cosmic chaos. And here I am, lucky enough to witness the result.

Diving down to get a closer look at a stromatolite, I think of the words of philosopher Alan Watts:

> Through our eyes, the universe is perceiving itself. Through our ears, the universe is listening to its harmonies. We are the witnesses through which the universe becomes conscious of its glory, of its magnificence.

Our consciousness is a marvellous and bizarre thing. Why, from the turmoil of the universe, has sentient life emerged?

There are probably more definitions of consciousness than there are fish in that super-salty pool. Well, maybe not quite, but there are a lot. As I swim, I experience how chilly the water is, I see the fish, and I have an incessant inner monologue dredging up fancy quotes from philosophers. All that comes together as an experience, it 'feels like something' to be me in this moment. This chimes with one of my favourite ways of thinking about consciousness, and the one I find most useful, which was fleshed out in an influential 1974 essay called 'What is it like to be a bat?' by philosopher and evolutionary thinker Thomas Nagel. This feels-like-ness is a useful way of thinking about consciousness.[1]

Watching the fish, I wonder what it is like to be them. What is their experience of this environment? Are they also seeing the parts of the universe perceiving itself? If so, what do they see? Feel? Am I scaring them? Or is it a sophisticated unconscious reflex that makes them flee from me?

Seeing them scatter and shimmer around the stromatolites, the next question I get to asking is, 'If these fish are conscious, when did the universe wake up? When did it start perceiving itself?' It turns out there is no scientific consensus as to which species are conscious, how consciousness evolved, or when it evolved. There are, however, a lot of strong opinions.

Some thinkers draw a line around humans as Earth's only conscious creatures, or humans and a select few super-smart animals like chimpanzees and dolphins. This would imply consciousness is only a recent development on Earth. Others suggest that all life has a spark of consciousness, so the bacteria that built the first stromatolites had some sort of awareness. But most evolutionary biologists believe consciousness is a property of animals; the most intense debate circles around which animals are conscious and why.

An analysis of 68 papers published between 2007 and 2017 described 29 theories of consciousness.[2] Discussing and analysing them all would take multiple books, so I will simply explore some of the expansive and restrictive views and contemplate what their implications might mean for us.

I consider the consciousness of many creatures in this chapter, but find myself returning repeatedly to fish. Like us, they are vertebrates, animals with backbones and internal skeletons. We all share a similar body plan and the same common ancient ancestor. In fact, the first vertebrates were simple, jawless proto-fish flitting about the ocean around 530 million years ago, maybe earlier.[3] Those proto-fish evolved into more recognisably fishy fish with jaws and rows of gills. A few of these species, which may have resembled today's lungfish, had two sets of tough limb fins, great for navigating the shallows and having sojourns on land. By 390 million years ago, this lineage evolved good walking skills, and these terrestrial adventurers would diversify into amphibians, then the reptiles, mammals and birds we know today. In that way, we are all fish, so fish don't really exist as a category ... But you know what I mean when I say 'fish'.[4]

Fish are where things get ugly in this scientific debate about consciousness. Fish also *seem* simpler, with smaller brains that do not have the bells and whistles ours do. Because they stayed in the ocean, it is tempting, but misleading, to argue that they remained 'primitive' while we continued to grow.

Why is fish sentience so controversial? There are two reasons. The first being the scientific implications of fish consciousness. Many hypotheses contend consciousness is dependent on physiological characteristics only mammals and birds possess, such as large brains with specific structures. If fish are conscious, it means these ideas need to be revisited and scientific acknowledgment of consciousness might have to extend beyond vertebrates to animals like octopuses, bees, or hermit crabs. Consciousness would also transform from being a relatively recent and rare occurrence to a more ancient and widespread one.

Second are the moral implications. This debate could conceivably centre around frogs or lizards but, while they are eaten by humans, neither are integral to human diets or exploited to the same degree that fish are. According to the Food and Agriculture Organization of the United Nations, an estimated 76.8 million tonnes of finfish were harvested by fisheries and 87.5 million tonnes of aquatic animals were raised in aquaculture in 2020.[5] Fish are a nutritionally crucial and culturally significant food source for many communities around the world. As current harvesting practices would likely cause extreme suffering to a conscious being, acknowledging their sentience can be uncomfortable to contemplate, setting us up for moral quandaries and discomfort.

The 'hard problem'

Understanding consciousness is often considered the 'hard problem' by scientists and philosophers alike. Why is it so hard? The idea that seemingly physical things, the brain and body, can produce a mind capable of thoughts, perception, feelings, desires and a sense of self, has been tripping us up for millennia. How we perceive the world through our senses and experience, and what order and meaning we apply to those perceptions, fascinated us well before the modern scientific method was born. The intellectual tradition around interrogating consciousness was particularly strong in ancient India, and views of what it is to be conscious deeply inform the Buddhist, Hindu, and Jain worldviews.

Reading translations of Dignāga, a fifth-century Buddhist logician and ponderer of perception, was a bizarre experience for me. I felt a portal had opened up before me and I was tumbling back in time – or maybe Dignāga was zooming into the modern world. His work on what we can know of the world through our perceptions reads as if it was written by the best modern neuroscientists, and indeed many modern researchers take inspiration from his texts. This is a testament to how enduring and intractable the hard problem is. Ancient Greek philosophers also tackled the question, their writings inspiring later works in the Christian and Islamic traditions.

Another challenge of the hard problem is that our experience is deeply subjective. We can each only truly inhabit our own individual consciousness, so rigorous scientific interrogation of the phenomena is tricky. We can never be sure of another human's experience, so being sure of another species' experience is even harder. This is called the problem of 'other minds'.

Over millennia countless solutions to the hard problem have been offered. Some contend that the 'mind' is a fundamentally separate thing to the body – that it kind of just hangs out in our noggin. Others contend that all matter is somehow conscious, and so as our bodies are made of slightly conscious stuff, it can manifest consciousness. This might sound a bit far-fetched, but it is argued by some serious physicists. However, I take a different view, in line with most, but not all, evolutionary biologists, that consciousness – or the mind – is literally the brain and its processes. This is called a materialistic or physicalist approach. But if you have a different perspective, don't worry – there is plenty in here for you.

The lights are on but nobody's home

How could a complex animal like a fish not be conscious? Fish have sophisticated sensory systems: they can see, sense water movement around them, hear, and more. So, they have data input. They can learn, so their brain can make associations and remember them. Many researchers believe there has to be a second step after data input to be conscious – a 'minimal inner awareness of one's ongoing mental functioning'.[6] Essentially someone must *be home* to observe that data. If they are not conscious, all this processing and sorting is occurring without it coming together as an experience. This view implies that fish are biological robots, existing in the world without being aware of it; that their world contains no pain, no fear, no joy.

So how could an animal as complicated as a fish not experience anything? It's not impossible – humans can have sensory experiences without being aware of them. Injury to certain parts of the brain can result in a phenomenon called blind sight, where a person will feel blind, or as if their vision is limited, but they are not. They can navigate around a room or point to various objects and describe them – though they cannot explain how they know what they are seeing, it feels like a hunch rather than an experience. This condition has been used to argue that data can go in and be processed without it 'feeling like something' to have that data processed.

Blind sight could be an example of consciousness faltering when that second step is missing, showing it is possible for information processing not to result in subjective experience. Most of these second step ideas argue that experience only comes with a very complex brain, and as these are relatively contemporary, the implication is that consciousness is rare and recent in evolutionary time.

Neuroscientists have identified which bits of the brain perceive things, like visual stimulus. If a second step is required, does it also rely on specific parts of the brain or actions of the brain? This was first argued by Francis Crick (yes, that Crick of the double helix) and Christof Koch of the Allen Institute. They proposed researchers identify the minimum things a brain needs for subjectivity to emerge.[7] These are termed the 'neural correlates of consciousness' and could include brain structures, numbers of neurons or patterns of brain activity.

To identify neural correlates of consciousness, human studies are often used to create a standard other species can be compared to. This approach has a big advantage – adult humans can tell us how they are experiencing their consciousness, so we can get reports of what it *feels like* when the brain is acting in a certain way or when areas of the brain are injured. As Günter Ehret and Raymond Romand put it in a 2022 paper:

> If neural markers in activity patterns of the human brain reliably reported about awareness and/or consciousness,

> this knowledge could be used as a new objectifiable gate for getting experimentally reproducible access to study the possible presence of awareness and consciousness in animals.[8]

Researchers can measure how a brain responds when attention is focused on certain things, like being asked to perform tasks in a lab, for example. The brain is an electrical organ. This is true on both a macro and micro level. Brain cells, or neurons, communicate by releasing and receiving chemicals and these messages are passed along through an electric charge. These individual pulses can synchronise in the brain, causing brain waves or, in science jargon, neural oscillations. The frequency and intensity of these waves differs depending on what we are doing. During deepest sleep they are low-frequency delta waves. During strong focus, the most intense high-frequency gamma waves are dominant. There are many intermediate points too.

Ehret and Romand single out focused attention as necessary for consciousness. This is because it represents processes taking that second step on from mere awareness of data. Attention allows for action informed by deliberation, the capacity to take note of how we are feeling, and possibly to have some level of control over thoughts and moods. As gamma waves, along with a few other brain activities, are associated with focused attention in humans, they highlight this pattern as a possible neural correlate of consciousness.

Gamma waves have been found in mammals, including rats, macaque monkeys and dolphins. Similar activity has also been found in the brains of crows, who are particularly intelligent birds. This suggests to Ehret and Romand that at least some of our furry and feathered friends could be capable of subjective experience. There is a problem with gamma waves, however. They are expensive. It takes a lot of energy to make your brain that buzzy. Things that are expensive generally get weeded out by evolution unless their benefits for survival outweigh their costs. This idea links to another prominent view in consciousness research – that only animals who can regulate their own temperatures, colloquially called warm blooded, can be conscious, or that consciousness is far more likely in warm-blooded (endothermic) animals.

Powering an internal heating system is expensive in terms of energy. To find the extra food needed to fuel the system, it helps if endotherms are smart. Neuroscientist Nicholas Humphrey argues that ectotherms – so-called cold-blooded animals like fish, lizards, bees and even octopuses that rely on the environment for their temperature – are not conscious and do not need to be. His argument is physiological as well as behavioural: the higher the body temperature, the faster the nerve cells move. The neurons of endotherms like us, whose bodies are usually kept well above 30 degrees C, have enough power to get the juices flowing for sentience. In a 2023 talk to the Royal Institute, Humphrey put it like this: 'Turn up the temperature and bingo, activity takes off and the phenomenal self emerges.'[9]

Ehret and Romand do not rule out fish, octopuses, or insects having consciousness. They suggest that a capacity to give selective attention should be considered. As the forms of selective attention they have identified in humans seem to be costly, researchers should consider whether subjectivity is an 'energetic luxury' a species in any given environment can afford and benefit from.

Another way to understand how consciousness emerges is to investigate the structure of the brain, rather than just its electrical goings-on. Brains in all species are organised affairs.

Vertebrates – fish, amphibians, reptiles, birds and mammals – all share a basic brain layout, although with very specific differences. During my zoology major I dissected a fish, slicing its head open to look at its brain. The mass of cells did not appear super 'brain like'; it looked rather small, elongated, and lumpy. Could much go on in there?

Despite being minimalist compared to our juicy brains, the fish brain is still a complex organ, divided into sections that have different roles. Like us they have a hypothalamus, which in humans keeps our biological processes running, contributes to mood, and manages hormones. They share other structures we have, like the cerebellum, which helps us with movement, among other things. The hippocampus and amygdala, which we use for long-term memory and fear respectively, are missing. However, they have analogous structures that seem to fulfil similar roles.

There is a big, pulsating difference, however. When I think of a human brain I picture the two giant cerebral hemispheres that look all … foldy. All that foldy stuff on the surface represents the cerebral cortex, and 90 per cent of the cerebral cortex is the neocortex, a part of the brain unique to mammals. This elaborate mass of cells, the cortical part of the brain, sits on top of the parts of the brain we share with fishes. We evolved this brain region after our evolutionary split with our shared ancestor with modern reptiles. Birds independently evolved a similar structure called the pallium, which shares many similarities with the neocortex.[10]

In humans, the neocortex is the 'thinking' part of our brain. It is where the language centres are, where we form speech, visualise things, plan, and exercise control. Injuries to a specific part of the cortex can often leave a person without the skill associated with that section, which occurs in blind sight cases. Many researchers of consciousness claim that the cerebral cortex, or specifically the neocortex, is the 'seat' of consciousness. Among these are some who believe that the whole cortex, or most of it, works to produce consciousness, or that specific areas of the cortex are where the keys to our inner life reside.

The assertion that consciousness is generated by the neocortex has been used to argue against fish consciousness. Invertebrate brains have evolved along a separate track. They are structured differently and tend to be less centralised than the mammalian cerebral cortex, and thus in this view they are also excluded from consciousness.

A cortical view that is enjoying particular popularity at the moment is the global workspace theory. Early work on this theory was done by neurobiologist Bernard Baars, and he and others have continued to refine the idea. It holds that different areas of the cortex are always unconsciously processing information to keep us ticking over.[11]

Think of it like this. Your brain may be processing the sound of a song playing in the background, a slight chill in the air or an ache in a muscle, all while not really being consciously aware of those things. While this is happening, you also have a system of long-term memory lying in wait in case you need that stored knowledge. Consciousness only appears when elements come into focus and connect. This coming together of information allows for a flexible mind that can make decisions about new problems.

While some proponents of the idea argue that fish or invertebrates may have the capacity to do this without a neocortex, most contend that a neocortex, and in some cases the prefrontal section of the cortex, is essential to get the job done.[12] These researchers generally reserve the title of 'conscious being' for humans and a handful of other animals, mainly primates.[13]

I am sceptical of this view. To bring it back to fish, behavioural researchers have shown that many fish have incredible cognitive abilities. They can create associations between unrelated things, such as colours or sounds, and food. They can then remember those associations for months or years.[14] Some fish, like rockpool-dwelling gobies, can create complex mental maps and make decisions based on those maps.[15] For gobies to navigate safely from rockpool to rockpool, they need to understand the special layout of their environment.[16]

The capacity to look in a mirror and know the face looking back is your own is held by many as a marker of self-awareness. Humans, chimpanzees and orangutans can do this, and contested research suggests dolphins, horses, elephants, magpies, and a crow can do it too. Multiple experiments have shown that cleaner wrasse, a fish that cleans parasites off other fish, also recognise themselves when presented with a mirror. When a dot is applied to it, and it is then shown its reflection, the wrasse will try to scrub the dot off on surrounding surfaces. When this research was first published it was met with scepticism, but further experiments have backed up the finding.[17]

My favourite example of fish intelligence is the alliance that forms between moray eels and groupers on coral reefs. Moray eels' slender bodies mean they can chase prey through the complex reef structure but they are slow in open water. Whereas groupers are speedy but cannot get into the reef structure, but will be ready to pounce when the eel flushes prey out into the open providing both hunters with a meal. Eels and groupers will form long-term alliances, showing the ability to recognise each other and navigate a social relationship.

Groupers will solicit eels to come hunting with them via a series of head shakes. The odd couple will then swim together, scanning for food. If prey disappears into the reef, the grouper will do a 'headstand' above where it disappeared – essentially 'pointing' at the spot the eel should focus on.[18]

This is complicated stuff and I believe it could indicate that both species are capable of some sort of theory of mind. The phenomenon that was first spotted in the wild is now being studied in the lab and has shown that the groupers are able to discriminate between competent eels that are efficient hunters, and incompetent ones.[19]

My concern about tying consciousness to a cortex is that it could lead us into a logical trap that goes like this: humans need cortexes for consciousness, therefore cortexes are needed for consciousness, therefore all animals without them have no consciousness. It is easy to make those leaps, and many have. Biomedical Scientist and pain researcher Brian Key of the University of Queensland writes, 'Fish lack a cerebral cortex or its homologue and hence cannot experience pain or fear.'[20] But there is an evidence gap in this statement that makes certainty on this issue hard for me to accept.

It is easy to assume that the fish alive today are static representations of the past. But their brains have also been evolving for hundreds of millions of years, and researchers who focus on them are beginning to realise that fish brains cannot be viewed as merely underdeveloped mammal brains. I worry that views presupposing the structures that support human cognition are needed for other species to perform complicated cognitive tasks risk missing alternative ways a brain could be organised.

Proponents of cortical models of consciousness are nervous about extending sentience far beyond humans. But Michael

Arbib of the University of Southern California closes the net even more tightly, claiming that only humans are conscious because: 'What makes human consciousness so different is that it includes expression of our thoughts in language'.[21]

He argues that early vertebrates, and modern vertebrates like frogs – and, one would assume, fishes – have no consciousness. Rather, they operate in a world of inputs and reflexes. Non-human mammals may have an 'animal awareness' of the world, where feelings such as fear, lust, or a maternal instinct may interact with their inputs and reflexes, with that system becoming more complex for primates such as monkeys. Our consciousness is made unique by language being laid over animal awareness, Arbib says: 'I suggest that language evolved for communication but as a corollary enriched our consciousness with an internal monologue.'

Some people do not have an internal monologue, and for those who do it can be experienced very differently. This has only recently come to scientific attention but periodically causes a stir on social media as the internet allows thousands of people to compare their inner lives.

I emailed Arbib and asked whether people without inner monologues disprove his hypothesis. He didn't think so. The argument isn't that you need to have an inner voice to be conscious, although his work does discuss its role. But that language, and the social skills you need to use it, changed the architecture of the brain in such a way 'that consciousness may seem to be sometimes observer and sometimes controller', as he puts it. Essentially, language shaped our brain so we could observe our surroundings and environments and structure our thoughts and reactions.

Humans, especially scientists, often enjoy defining themselves by their intelligence and reason. In many ways, humans are our best shot at understanding consciousness because we can ask each other to describe what it 'feels like' to be the other person in any given scenario. But we are just one species, and a weird one at that. So, inferring the lack of consciousness of other species, especially very distantly related ones, runs the risk of defining alternative consciousness out of existence.

Feelings before facts? I feel therefore I am

Human brains can certainly engage in incredible feats of cognition; we are smart. But so are computers. Information processing and problem solving exist in those non-living systems. But currently most of us do not consider our tech to be conscious. What then, is the source of consciousness? Neuropsychologist and psychoanalyst Mark Solms argues that if we are trying to figure out what it 'feels like' to be something, then feelings are a good place to start. Because, as he says, 'You cannot have a feeling that you do not feel.' He lays out this argument in his book *The Hidden Spring*, which contends that subjective feelings like hunger, thirst, pain and pleasure are the foundations upon which all consciousness stands.

This view might seem woo-woo compared to the hard logic employed by global workspace theoreticians. But Solms argues that feelings are serious business and what animals use to stay alive. This is because feelings are an extended process that keeps our bodies in a stable state called homeostasis. Sweating to cool down as a response to feeling hot is a type of homeostasis. But sweating might not be enough to beat the heat. Solms' extended version would be making the decision to seek shade, have a cold drink or go for a swim. These behaviours are voluntary add-ons motivated by the subjective unpleasant experience of overheating. The feeling is needed to kickstart the thought and initiation process that leads to behaviours that cool us down. His argument is supported by complicated maths and physics that he developed with Karl Friston of University College London but, as this is a book about evolution, not physics, we won't get into the details.[22]

Researchers proposing cortical theories of consciousness base their models on humans, and the same is true for Solms. However, he rejects cortical theories of consciousness outright, describing them as 'cortical fallacies'. Many of his patients have had injuries to sections of the cortex, and in his book he details how, while they might have specific issues with skills like memory retrieval, these patients are still conscious. However, his most compelling piece of evidence is that some people are born without cerebral hemispheres, and thus have no cortex. This is a condition called hydranencephaly, and people who have it seem to be conscious.

For decades people with hydranencephaly were assumed by many in the scientific and medical establishments not to be conscious. This was despite parents and carers believing them so. While people with hydranencephaly are dramatically limited in their abilities, not having the 'thinking part' of the brain, they do appear to have feelings: they will laugh, cry, have preferences, show pleasure, and can engage with simple sequences of play.[23] The rejection of hydranencephaly patients' sentience severely impacted how they were treated by the medical system. Solms is inspired by the work of neuroscientist Bjorn Merker, who has worked with hydranencephalic children and their carers and is now an impassioned advocate for the sentience of these patients to be acknowledged.

Solms and Merker believe the consciousness in hydranencephalic children is further supported by animal experiments. Rats that have had their cortexes surgically removed in the lab remain behaviourally complex. They play, groom themselves, avoid pain and do all the things you would expect a rat to do. This extends to mating and rearing young, although some of the behaviours they display in pursuing these goals are 'abnormally executed'.[24] For these reasons and more, Merker has pushed for a view of consciousness that does not require the cortex, arguing that the 'brainstem mechanisms [the bit we share with fish] are integral to the constitution of the conscious state'.[25]

Despite dismissing the cortex as the seat of consciousness, Solms is also in pursuit of neural correlates of consciousness. He, like Merker, highlights the brainstem, specifically a section of it called the reticular activating system, as the minimum thing a brain needs for subjectivity to emerge. Why? This area controls our feelings. Small injuries to this part of the brain via lesion or stroke can, as Solms puts it, 'obliterate' consciousness. Solms makes the provocative argument that: 'Cortical processes are fundamentally unconscious things, they are simply algorithms if left to their own devices. Consciousness, all of it, comes from the brainstem.'

As this part of the brain is shared by all vertebrates, Solms has no problem extending consciousness to fish. This is backed up by the fact that vertebrates seem to seek pleasure and avoid things they do not like. Research does seem to confirm that fish have feelings. A 2017 study of emotional states in sea bream showed that they are capable of highly stimulated positive affect, low-level positive moods, highly stimulated negative moods, low-level negative moods. In humans these could translate to happy, relaxed, fearful and sad.[26]

In humans, the brain chemical oxytocin helps us to feel the emotions we perceive or imagine in others – in other words, to be empathetic. The same seems to be true in fish. In 2023, when researchers knocked out oxytocin genes in zebra fish, the animals became loners and did not respond if other fish became agitated. When the mutants were injected with oxytocin and observed other fish acting in a way the researchers perceived as 'fearful', they reacted similarly to their scared tank mates. This suggests that emotions are not only present in fish, but complex and occur in a rich social context.[27]

Solms suspects that invertebrates such as octopuses and crustaceans are conscious, but is less certain of it, as they do not have a brainstem. If his view that feelings evolved to support extended homeostasis can be supported, I believe this would imply that any animal that has the capacity to change its circumstances through behaviour would have some sort of consciousness.

There is strong evidence for feelings in invertebrates. A startling example comes from bee research that suggests bees can have good or bad moods, even getting mini bee depression. Bees have tiny but highly interconnected brains and are capable of learning; some researchers even claim they play.

A 2011 study even showed that individual bees seem to get into depressive funks.[28] British ethnologist Melissa Bateson and her team treated some bees to a great time, letting them relax and consume a sugary liquid. Bliss. Other bees they shook in a vial, something that doesn't hurt them but they seem to hate. Afterwards both sets of bees were exposed to a substance they had not experienced before. The blissed-out bees were curious and investigated, whereas the shaken-up bees were wary. The researchers suggested their negative experience made them more pessimistic. When analysed, the blissed bees had higher levels of dopamine, the brain chemical involved with reward. We humans show this exact pattern: when we are feeling low, we are more likely to predict negative outcomes than when we are happy.

The cortical crowd are cynical of evidence pointing to fish or invertebrates having feelings. James D. Rose of the University of Wyoming, who is a strong advocate for cortical consciousness, argues that emotions, or the processes that cause us to have them, do not need to be experienced to serve a function, writing:

> Emotions are autonomous and functional in their own right, yet they also provide the pre-conscious raw material for the experience of conscious feelings, which arise through further processing by higher cortical regions. These cortical regions are essentially the ones that underlie the conscious experience of suffering in pain.[29]

In this view, the chemical aspects of emotion are divorced from experience. I struggle with this. I'm inclined to think that the capacity to feel the affective experience of emotion was what was being selected for when the complicated mechanisms that create emotion evolved. In other words, what's the point of evolving a feeling if you can't feel it?

Rose disagrees with the idea that complex behaviour in vertebrates without a cortex implies consciousness. To him, the research on cortex-free rats has a very different implication than it does to Solms. He suggests it illustrates just how much complexity can exist without the need for consciousness. Rose views behaviours such as those the rats displayed as complicated pre-programmed 'stereotyped' behaviours that do not require consciousness. To him, this research shows that even when fish have complicated behaviours and 'emotions' can be detected, this does not imply that they can bring these data points together and have a subjective experience.[30]

A focus on feelings as the first, or at least early, spark of consciousness opens new cognitive views of sentience that are expansive and suggest conscious life has an ancient history. This is the case for the unlimited associative learning model of consciousness, which sees a role for feeling and cognition. It was developed by evolutionary thinkers Eva Jablonka of Tel Aviv University and Simona Ginsburg, and unveiled over a series of years in the 2000s resulting in a 2019 book *The Evolution of the Sensitive Soul: Learning and the Origins of Consciousness.*[31] They argue that a species can be said to be conscious if it can perform a certain type of learning.

To understand, let's explore a real-world example. It is your anniversary and time to find your special someone the perfect present. You drive to town and feel excited because there is a park right where you need to shop, making this shopping strip a great option. You go into a swanky bag shop you have never visited before and ask the assistant for help. They look you up and down, say something judgy and are generally rude. You leave with no bag and feel embarrassed. That bad feeling teaches you not to like that shop and you buy perfume instead. You have made an association between that shop and feeling bad and decide never to go back. Later that night, at your fancy dinner date, you try crème brûlée for the first time, it tastes delicious, and you learn you like crème brûlée. A few weeks later you find out from your friend at the perfume shop that the rude assistant in the bag shop was fired because the establishment has a policy of making their customers feel welcome. When you hear this, you update your prior opinions, shop there again and have a lovely experience. From now on you will return there to buy presents.

This is a micro version of unlimited associative learning. You were able to learn by making an association between something, the bag shop or the dessert, and a feeling of embarrassment or pleasure. You can also link associations. The great car spot is near the two shops you now like. Crucially, you can also make multiple unrelated associations, such as your crème brûlée discovery. You were also able to change your mind about the bag shop after having a good experience.

For this type of learning to exist, feelings must be present. Humans can learn abstract facts – for example, how evolution by natural selection works – but to learn how to operate in the world, feelings are essential guides. If you are a fish foraging on a reef and you learn that A is bad and B is good and thereafter avoid A, 'good' and 'bad' must have some intrinsic property – it must 'feel like' something. To me, these subjective judgments about the quality of things allow for consciousness to pair with cognition.

Who is conscious under this view? Making the capacity for unlimited associated learning the threshold for consciousness means that life on Earth 'woke up' independently in three separate lineages, according to Ginsburg and Jablonka. These were the early vertebrates, early crustaceans and the mollusc line that would lead to cephalopods such as octopus and squid. They believe this happened during the Cambrian explosion, an exciting million-year period around 530 million years ago when the relatively slow-moving, not super-diverse life forms Earth had previously hosted suddenly exploded into an eclectic array of new and mobile ocean-dwelling species. As species' sentience winked into existence, this created an 'arms race' as the new traits of improved perception, attention, learning and feeling provided some individuals with advantages over others.

I am more convinced by explanations that centre feelings and presume consciousness is common among animals than I am by restrictive views that require specific structures or processes. Systems that allow us to have positive and negative experiences seem to be more ancient than the systems that organise those feelings into rational thought. In an evolutionary context this makes more sense to me. Why? Because if consciousness sparks into being only in very complex animals, that implies there is a lot of learning going on without an innate 'feeling system' guiding it, and that seems unlikely.

Hard problems or hard hearts?

There is no agreement on where or how consciousness starts. So, while I lean towards an expansive view on consciousness being an ancient attribute, I do not think the theorists who have a restrictive view of consciousness are at fault. The science simply is not settled, and with over 2500 years of recorded disagreement on the topic, I cannot imagine there will be consensus any time soon. Given our uncertainty as to who and what is conscious, I do worry when restrictive ideas are hauled in as facts and used to determine the way other individuals are treated – we tend to care a lot more about the treatment of conscious beings than non-conscious entities.

Pain is the body's response to damage. In vertebrates, particular sensors called nociceptors receive information that something is not quite right and send that signal to the brain, which then generates our experience of pain. Although this can misfire when pain is experienced without damage. Pain is often identified in animals when they try to protect an injured part of their bodies, or when they seek out pain relief like opioids. Pain, or at least the perception of physical damage, is thought to have evolved as animals who experience pain might be less likely to injure themselves or more likely to rest and heal.

Consciousness is considered necessary to feel pain, but pain is not necessary for consciousness.

Views that are sceptical of others' consciousness have led to cruelty towards human babies and marginalised adults, including on the operating table. This is tough going, so make a cuppa, take time for yourself and if it's really not the day for you, skip to the next section.

Until the 1980s it was not uncommon for surgery to be performed on babies without anaesthesia or pain relief. I did not believe this when I first read it, it is so horrific. How could it be true? How could scientists and doctors who presumably had their patients' best interests at heart do this? At the time doctors believed now-disproved research that denied infant pain and, often, infant consciousness.

Denial of infant pain seems to have lurched into the academic world in the past 200 years, with several studies in the early part of the 20th century concluding infants felt no pain. One of the most influential studies propping up this belief was undertaken by developmental psychologist Myrtle McGraw and published in 1941.[32] In this study she pricked 75 infants and children with sterilised blunt safety pins from right after birth to four years of age, making a total of 2008 observations. In the first few hours or days of life, a prick would induce 'diffuse bodily movements accompanied by crying'. Older babies tried to 'push the pin away', showing they knew where the sensation was coming from.

Reading the results of this paper I was shocked it could ever have been used to argue against infant pain, as it detailed significant reactions to the pricks from birth. McGraw herself does not make this argument. The key finding of her paper was that an infant's reaction to pinpricks changes with age, becoming more specific from the second or third year of life. However, she does link her findings to scientific beliefs at the time that 'no part of the cerebral cortex is functioning appreciably at birth'. Extrapolating from the view of the day, she reflects that it is plausible that 'experiences of the newborn infant do not extend beyond the subcortical or thalamic level'. From this observation she posits this may mean that babies have diminished sensory experience.

McGraw's speculations were picked up and run with. In a 2013 paper describing the dishonourable history of pain research, psychologist Elissa Rodkey of Crandall University and early life pain researcher Pillai Riddell of York University write that McGraw's 'work would be frequently cited by later infant pain researchers and her tentative conclusions about brain development used as statements of fact'.[33] This included research that made far stronger claims, most referencing the cortex, that denied infant pain. Here we run into ideas of cortical consciousness again, where researchers are hesitant to infer consciousness if they perceive an individual's cortex is not up to muster. Rodkey and Riddell note:

> ... even though many of these studies were inconclusive, since scepticism about infant pain was more in tune with the widespread mechanistic views of the infant, they were eventually interpreted into scientific consensus.[34]

Even while surgeons were confidently denying babies felt pain, there were scientific dissenters whose research contradicted the dominant paradigm. But it took the combination of activism by parents and a high-profile 1987 paper by paediatrician Kanwaljeet Anand and anaesthesiologist Paul Hickey for the reality of infant pain to be accepted.[35] Since then, study after study has supported the existence of infant pain. But what about the cortex? It turns out early researchers underestimated the power of a baby's cortex, and while it does function a little differently to that of an adult, it is indeed up and running. A 2015 study by Sezgi Goksan of University College London and colleagues scanned the brains of newborns while the bubs were given a small pain stimulus.[36] It showed that the cortexes of babies' brains respond in the same way an adult's does.

Babies are not the only humans to have been denied the dignity of their pain being recognised. In the 19th century, American physician Samuel Cartright argued that enslaved black people were 'insensible to pain when subjected to punishment', a self-serving view from a rabidly pro-slavery thinker.[37] Similar views led to great atrocities, including experimental surgery with little sedation, on people of African descent in the United States. While the science is in that there is no difference between people of different racialised identities, this view still impacts patients today. A 2016 study of medical students in the United States led by social and behavioural scientist Kelly Hoffman found that around half the white students surveyed carried beliefs about black people and pain similar to those propagated during the time of chattel slavery. Disturbingly, students with those beliefs rated black patients' pain as lower than that of white patients.[38]

Misunderstanding and active misrepresentation of evolutionary theory was used to justify these views of pain and consciousness. Views of evolution that see natural selection as a ladder of superiority reaching its peak in humans – and for racists and other bigots, in white non-disabled men – have used this view of life to devalue the experiences and moral value of others. Rodkey and Riddell write that in the late 1800s, 'Children were lumped together with animals, savages, and the insane as primitives whose emotional expression was simply reflex actions reinforced by habit.' Non-white people and people with disabilities got short shrift here too. Scientific racism that positioned non-white people as less rational, and thus less feeling, was used to justify enslavement, colonisation, and genocide. Similar views resulted in mass sterilisation of disabled people.

These hierarchal views of evolution are deeply misleading; everything alive on Earth today is equally evolved.

Fish feelings

Now back to fish. It feels a bit jarring to go from discussing the suffering of babies and marginalised people to non-human animals. But most of us see infants as precious, wonderful, vulnerable, and crucial to protect, including, presumably the scientists who denied their pain. As such, I believe the story of how science tricked itself into enormous cruelty against babies should serve as a warning. The experience of non-human animals, especially ones as different from us as fish, is alien. If these creatures are not conscious, it implies that the decisions we make when we interact with them do not need to consider their welfare. However, if fish – and by extension crustaceans, cephalopods and insects – are conscious, their welfare should be an active factor in how we choose to interact with them and the environments they live in. So, is science at risk of repeating its mistakes? Especially as we all stand to gain from denying fish sentience.

In the great fish debate, some argue that we should err on the side of caution and assume sentience until such time as there can be no doubt fish are not sentient. If we do not do this, we could accidentally cause immeasurable suffering for fish – and other species like lobsters and octopuses – that could have been avoided. Others argue the opposite. An extensive 2023 article by Ben Diggles of DigsFish Services, and his academic colleagues from universities from multiple countries, strongly suggest we should assume fish are not sentient.[39]

Why dismiss fish feelings? The Diggles and colleagues, school of thought is sceptical of the science of fish sentience. This comes from a reasonable concern that too much focus on fish welfare might radically restructure our food systems and threaten global hunger – not an outcome anyone would want. They also argue that if an acceptance of fish sentience influences the laws dictating how they can be treated, this might devalue the concept of animal welfare. Essentially, if fish are lumped in as sentient with animals like cows and pigs, we may care less for all of them. Both are valid concerns, however, I am not sure that seeing fish and possibly invertebrates as sentient would grind our food systems to a halt, though it might fuel motivation to limit suffering. Pain is an inevitable part of life, and if we are making decisions for another creature, even if that involves them feeling pain, it is better to know about it.

A critique often levelled at people whose science leads them to expand the definition of consciousness is that they are projecting human experiences onto animals. This is called anthropomorphism and is considered a cardinal sin in zoology. In a paper denying fish pain, James Rose and colleagues argues:

> Anthropomorphic thinking, a bias that obstructs objectivity about biological questions, would lead to expectations that fishes should be more like humans in their capacity for nociception, even pain.[40]

True enough.

I have certainly been guilty of fish anthropomorphism. When I was a kid, I used to go spearfishing with my dad. It was great fun. But at around age 11 or 12 I started to get uncomfortable. A few times I made bad shots, spearing a fish through the body, not the head. I would hold onto my spear and watch the fish thrash around before dying. I gave up spearfishing, despite loving it, and my dad followed suit.

My feeling of moral horror watching a fish struggle on the end of a spear is not evidence that fish experience pain, however, the intensity of my reaction demonstrates how, for softies like me, it's tempting to unscientifically project subjective knowledge of pain onto the fish's experience.

Equally, our fear that we might be perceived as unscientific and not rational can flip our scientific minds into an equally irrational space, where we deny feeling to anything too strange. The ultimate unknowability of other minds is not a justification for dismissing their existence. A fear of anthropomorphism is fuelling a reticence to grant consciousness to beings whose lives are manifestly different to ours. This falls into another trap, that of anthropocentrism, where the definition of consciousness is so centred around human experience that different ways of being are dismissed.[41]

I think the solution to this problem is expressed well by Thomas Nagel in his paper 'What is it like to be a bat?', where he writes: 'My realism about the subjective domain in all its forms implies a belief in the existence of facts beyond the reach of human concepts.'[42] In other words, as uncomfortable as it might be, we need to accept and respect realities that are fundamentally unknowable to us.

An important question to grapple with if we are to reach any kind of agreement as to who is conscious is, *why* we are conscious. Does sentience serve a purpose? Unsurprisingly, evolutionary biologists are divided on this question. Some believe that consciousness is a trait that can aid survival. Therefore, as consciousness arises, it is specifically selected for.

Others believe it is a by-product or 'spandrel'. Consciousness being a spandrel is implied by the unlimited associative learning view of consciousness, where traits like the ability to discriminate between good and bad, to have focused attention, and to learn through associations are selected for, and combined get a creature over the threshold for consciousness.[43]

Maybe our view of consciousness is too simplistic? The way I have been writing about it so far implies that either:

(a) There is a threshold that is met, then *BAM*, an animal is conscious. Or:
(b) Consciousness is on a spectrum from minimal consciousness to full-throttle sentience.

As we have seen, hierarchies are rarely useful when trying to understand the world from an evolutionary perspective. A principal investigator on the Foundations of Animal Sentience Project, Jonathan Birch of the London School of Economics, and colleagues, have argued for a perspective that busts the spectrum. They write:

> If we try to make sense of variation across the animal kingdom using a single sliding scale, ranking species as 'more conscious' or 'less conscious' than others, we will inevitably neglect important dimensions of variation.[44]

Birch and colleagues suggest that we view consciousness on a multidimensional framework. Their framework maps dimensions of consciousness onto shapes, say a hexagon, where each corner is 'dimensions'. Depending on how many elements of consciousness are being evaluated, the shape could be expanded to any cornered shape such as an octagon or decagon, or reduced to a square. The dimensions represent capacities such as:

- Perception, like sight or smell.
- Capacity to make evaluations through pleasure, pain and other feelings.
- A sense of being a 'self'.
- An integration of experience through time.

These dimensions can then be mapped onto a shape for a species. The stronger a species' capacity for a particular dimension, the closer the line goes to that dimension corner. As Birch and colleagues write, this 'allows the conscious states of animals to vary continuously along many different dimensions, so that a species has its own distinctive consciousness profile'.

A tool like this gives a more granular way to understand how the elements of consciousness might exist or be selected for within a specific species. For instance, let's compare my mini schnauzer Sadie with a cleaner wrasse and myself, and create a triangle framework that compares three dimensions. The dimensions are vision (for a perceptual example), emotions, and sense of self as judged by being able to recognise oneself in a mirror.

Sadie does not have great vision, so the cleaner wrasse and I would beat her on that account. However, judging by her ecstatic hellos, terror at the vet and love of treats, I suspect she feels her emotions more strongly than I do. I am not sure about the emotional state of a wrasse, so let's assume, for this thought experiment, it feels less strongly than I do. Sadie cannot recognise herself in a mirror, however, contested – yet compelling – research shows the wrasse can, and I can. On this test of consciousness, mapped onto a triangle, we all get different wacky triangles hinting at our unique ways of operating in the world.

This approach opens new, exciting avenues from an evolutionary perspective. Humans are restricted in our capacity to understand how other species experience their existence. However much we love our dogs, we can only guess what it is like for them to go for a walk and experience the world through their far superior sense of smell. But an evolutionary perspective means that we are not the standard that consciousness is measured against. Instead, it can be understood as a trait, or collection of traits, that have evolved and then been selected for in an array of environments over time, leading to different, specific experiences. In this way the history of human experience and consciousness is just one of many histories, and it does not have to be the standard other species must meet to be worthy of the label 'conscious'.

Could evolutionary approaches extend the bubble of consciousness beyond animals? Anthony Trewavas of the University of Edinburgh is among those who make the claim that plants are conscious, although he sometimes takes a linguistic view of consciousness, claiming that all non-human sentience is 'awareness'.[45] In a letter published in the journal *Trends in Plant Science*, Trewavas and colleagues argue against an animal-centric approach to consciousness. In fact, they open up space for plant consciousness being even more unknowable than, say, that of a fish. They write that, compared to animals, 'plant consciousness appears to be a different "breed" altogether'. If we accept plant consciousness, which is still a marginal view, it opens up tantalising questions about the evolution of subjective experience. The authors express this, writing that 'paying due respect to evolutionary considerations … brings the idea of plant consciousness to the fore'.[46]

Could 'other minds' exist without brains?

I have been staying firmly in the animal realm. After all, it was in our often mobile and behaviourally complex kingdom that the brain and nervous system evolved. While challenging the idea that we need a certain kind of brain for consciousness, I have been implying a need for a brain. But what about plants, fungi and single cells? Unlimited associative learning does not necessitate a brain, and while Mark Solms' 'feelings first' work centres around vertebrates, his logic does not. Could consciousness have evolved in other kingdoms of life?

In 2017 plant intelligence researcher Monica Gagliano of Southern Cross University took the leap and made the claim that plants are conscious, writing: 'Plants too must evaluate their world subjectively and use their own experiences and feelings as functional states that motivate their choices.' Gagliano came to this view after a series of experiments on *Mimosa pudica*, otherwise known as the 'sensitive plant'.[47] The species gets its name because, after sensing a disturbance, it quickly folds its leaves together, making it harder for predators to eat it. When its leaves are folded its capacity to photosynthesise, something it needs to do to make its own food, is reduced by 40 per cent. Gagliano wanted to see if it could learn to be habituated to false threats.

Gagliano and her team tested this by dropping potted sensitive plants from a set height. The plants did not appreciate this, and duly folded their leaves. While dramatic, this drop did not correlate with damage to the plants. Could they learn that, and stop folding their leaves when they were dropped? They certainly could, and fast. The researchers had suspected that if the plants could learn, it would take a while, so planned a training regime of 60 drops. The plants played by their own rules and learned to keep their leaves open within just a few drops. To check that the plants weren't just getting worn out, the researchers added another stimulus, like shaking, to see if the leaves were still reactive. They were, closing after a shake but remaining open after a drop.

Sensitive plants have remarkable memories. Even after 28 days without being dropped they would remember that a fall did not mean danger. Fascinatingly, they seemed to contextualise these lessons. If a plant had had it easy, with lots of access to light, it would still close its leaves after the first few drops even after a month-long break, but it would take fewer repeats to get back to not responding. Plants that had done it tough, under lower light conditions, had more to lose (time to photosynthesise) and tended to stay open from the first drop even a month later. Gagliano expanded her research and demonstrated that pea plants were also able to learn by association.[48] The plant scientist's perspective is very similar to Mark Solm's 'feelings first' approach, as she claims it is feelings that allow plants to make associations and learn.

These are amazing skills, but could they be happening without it 'feeling like' something to be a plant? As we have encountered, many consciousness researchers will deny consciousness to animals such as fish on the basis that they do not have a cortex, the right brainwaves or the capacity to bring their experiences together in a 'global workspace'. Those theoreticians would, of course, dismiss Gagliano's argument.

But consciousness researcher and author Jon Mallett of the University of Washington and colleagues dismiss the argument on different grounds. They claim that Gagliano's work shows evidence for the plants 'learning' an automatic reflexive response to a stimulus.[49] However, for consciousness to be present they need to display a behavioural response. A classic example of this difference can be made between dogs who have been taught to associate the sound of a bell with food and salivating compared with dogs being taught to push a button to receive food. This makes me think of Solm's view that consciousness is 'extended' homeostasis. It would be interesting to see if these associations are automatic, like sweating, or behavioural, like seeking shade. Though untangling the difference between behavioural and automatic in plants seems like a near impossible task.

In animals, the nervous system allows rapid electrical signalling across the body and in the brain. There is growing evidence that plants also use electricity to signal, but in a more diffuse way than nerves – though this comparison is contested, as the charge is small and similar types of electrical activity exist in animals outside their nervous systems. Fascinatingly, plants also use something called gamma-aminobutyric acid (GABA) for signalling. In animals, GABA is considered a 'primary inhibitory neurotransmitter' in the brain, but exactly how it works in plants is a little mysterious. A 2023 study led by Yuri Aratani, Takuya Uemura and colleagues from Japan's Saitama University tracked where the message was received and how it spread through *Arabidopsis*, a relative of cabbage.[50]

Plants are in constant dialogue with each other. This communication happens via chemicals released into the air, via underground networks built by fungi, and possibly through sound. After being damaged, say by being chomped by a caterpillar, a plant will release volatile organic compounds into the air. Other plants will then sense these chemicals and prepare their defences. This response has been found in at least 30 plants to date. The response is fast. In under 20 minutes the whole plant had the message, showing just how speedy plant communication and reactions can be. Many plant species can also sense when they are related to others, competing more with non-relatives than they do with their genetic kin.

New research on plant communication, signalling and learning is evidence of complexity, not evidence of consciousness. As Solms points out, computers can be complex without being conscious. The plant consciousness debate is being paralleled by similar debates concerning the sentience of fungi, strange aggregations called slime moulds, single cells and entire connected forests. I think it would take a huge amount of evidence to convince most scientists, and me, that non-animals are conscious. But this does not mean the research is on a hiding to nothing. The people asking these questions have revealed amazing things about the evolved complexity of life on Earth, things we may not have discovered otherwise. It also challenges us to expand our thinking and interrogate assumptions we make when contemplating how another entity experiences the world.

From our ancestors

This book has tracked evolutionary stories from life on Earth. I have been drawn to questions that touch most of our lives. Each question also has an element of tension at its centre, something that at first glance seems paradoxical. But what I return to in the quieter moments is, why are we here at all? Why, from the turmoil of the universe, has conscious life emerged?

These thoughts were especially present on my research trip to visit Gutharragudu/Shark Bay to see the stromatolites. My mother, Jackie, came on this trip with me and before flying north we had a few days in Perth to see stromatolite fossils at the Western Australian Museum (Boola Bardip), where a 3.4 billion-year-old specimen is displayed. It is at the far end of a dimly lit exhibition space illuminated with the same reverence given to religious relics in churches – this respect is well deserved as it is one of the oldest fossils ever found.

It is long, narrow and a striking orange colour. The label describing it explains that the evidence of its origin can be found in the 'egg carton' pattern of the rock. Shapes like this could not be made by processes such as wind, water, or erosion. These stromatolites were thought to have grown in the warm mineral-rich pools at the base of an ancient volcano. Those were the good old days, when things were simple. There was no war, no pain. In fact, there were no animals, no multicellular organisms at all. OK, so maybe they were the bad old days.

As Jackie and I glide through the crystal blue water enveloping stromatolites made by bacteria over thousands of years, my mind keeps drifting back to the 3.4 billion-year-old fossil in Perth. The tiny microbes making the stone structures below me are so like their forebears that made the fossil in that rock. The study of evolution is the study of ancestors, how the lives of those who came before inform what exists now. In the few days we spend in Gutharragudu, we are awed by life: we see sharks, turtles, dugong, a manta ray, corals, emus, little songbirds, and more. Suddenly, the idea that all that beauty has come from such simplicity hits me.

At one point I pop my head up to check on Jackie, who has stopped swimming. 'You OK?' I call out. I think she is in trouble in the water, but she is just struck by the power of the place, 'I feel quite emotional,' she responds. Neither of us can wipe the grins off our faces – a frustrating problem when snorkelling – the grandeur of the place has swept us away. In that moment we are lost in awe that the riotous abundance of life on Earth could emerge from cells so small.

When I wonder why I am here, I consider myself as a thinking, feeling descendant of the millions of generations that came before me. This view of life may not be shared by everyone, and is my philosophy, not my science. But I think that evolutionary history has primed us with an extraordinarily complex mind, one that has tendencies to great kindness and tendencies to great cruelty. The way I see it, we are on this planet, we are sentient, and with that we can choose to improve life for all its denizens.

Notes

Introduction

1 Allen, JF. (2016).'A proposal for formation of archaean stromatolites before the advent of oxygenic photosynthesis.' *Frontiers in Microbiology* 7 (2016): 1784.

2 Demoulin, CF et al. (2019), 'Cyanobacteria evolution: Insight from the fossil record.' *Free Radical Biology and Medicine* 140 (2019): 206–223.

Chapter 1 Why do we care?

1 Dawkins, R. (2006*). The Selfish Gene: 30th Anniversary Edition with a new Introduction by the Author*. Oxford: Oxford University Press.

2 Hamilton, W. D. (1964). 'The genetical evolution of social behaviour'. II. *Journal of Theoretical Biology*, *7*(1), 17–52. <doi.org/10.1016/0022-5193(64)90039-6>

3 Allen, B, Nowak, MA & Wilson, EO. (2013). 'Limitations of inclusive fitness.' *Proceedings of the National Academy of Sciences*, 110(50), 20135–20139.

4 Maynard Smith, J. (1997). 'John Maynard Smith – The point of evolutionary game theory.' (R. Dawkins, interviewer). <www.youtube.com/watch?v=FKegQW_lsGI>

5 Trivers, RL. (1971). 'The evolution of reciprocal altruism.' *The Quarterly Review of Biology* 46.1 (1971): 35–57.

6 Augusto, JF, Frasier, TR & Whitehead, H. (2017). 'Social structure of long-finned pilot whales (*Globicephala melas*) off northern Cape Breton Island, Nova Scotia.' *Behaviour* 154.5 (2017): 509–540.

7 Hersh, TA et al. (2022). 'Evidence from sperm whale clans of symbolic marking in non-human cultures.' *Proceedings of the National Academy of Sciences* 119.37 (2022): e2201692119. Whitehead, H. (2024). Sperm whale clans and human societies. *Royal Society Open Science*, 11(1), 231353. <royalsocietypublishing.org/doi/full/10.1098/rsos.231353>

8 Heide-Jørgensen, MP, Bloch, D, Stefansson, E, Mikkelsen, B, Ofstad, LH et al. (2002). 'Diving behaviour of long-finned pilot whales *Globicephala melas* around the Faroe Islands.' *Wildlife Biology*, 8(4), 307–313, (1 December 2002).

9 Irvine, L et al. (2017). 'Sperm whale dive behavior characteristics derived from intermediate-duration archival tag data.' *Ecology and Evolution* 7.19 (2017): 7822–7837.

10 Augusto, JF, Frasier, TR & Whitehead, H. (2017). 'Characterizing alloparental care in the pilot whale (*Globicephala melas*) population that summers off Cape Breton, Nova Scotia, Canada.' *Marine Mammal Science* 33.2 (2017): 440–456.

11 Teunissen, N, Fan, M, Roast, MJ, Hidalgo Aranzamendi, N, Kingma, SA & Peters, A. (2023). 'Best of both worlds? Helpers in a cooperative fairy-wren assist most to breeding pairs that comprise a potential mate and a relative.' *R. Soc. Open Sci.*10231342231342.

12 Augusto, JF, Frasier, TR & Whitehead, H. (2017). 440–456.

13 Gero, S, Engelhaupt, D & Whitehead, H. (2008). 'Heterogeneous social associations within a sperm whale, *Physeter macrocephalus*, unit reflect pairwise relatedness.' *Behavioral Ecology and Sociobiology*, 63, 143–151.

14 Oremus, M, Gales, R, Kettles, H, & Baker, CS. (2013). 'Genetic evidence of multiple matrilines and spatial disruption of kinship bonds in mass strandings of long-finned pilot whales, Globicephala melas.' *Journal of Heredity*, 104(3), 301–311.

15 Kean, Z, & Mitchell, N. (2019, December 1). 'Selfish by nature? Two scientific renegades who looked for kindness and paid a price' [Radio documentary]. ABC Radio National. <www.abc.net.au/listen/programs/sciencefriction/selfish-by-nature-or-not/11729916>

16 EO Wilson famously called Dawkins a journalist when critiquing his adherence to the gene's eye view. As a journalist it is a funny insult to me, but I guess it's not something a 'serious scientific mind' wants to be called.

17 Cantor, M, Shoemaker, L, Cabral, R et al. (2015). 'Multilevel animal societies can emerge from cultural transmission.' *Nat Commun* 6, 8091 (2015). <doi.org/10.1038/ncomms9091>

18 Trivers, R. (2020, May 1). Blog 'Trivers on Epstein. <roberttrivers.com/Blog/Entries/2020/1/5_Trivers_on_Epstein.html>

19 The Human Generosity Project. (2023). Karimojong | Human Generosity Project. <www.humangenerosity.org/field-work/karimojong>

20 Aktipis, A et al. (2018). 'Understanding cooperation through fitness interdependence.' *Nature Human Behaviour* 2.7 (2018): 429–431.

21 Ibbotson, P, Jimenez-Romero, C & Page, KM. (2022). 'Dying to cooperate: the role of environmental harshness in human collaboration.' *Behavioral Ecology*, Volume 33, Issue 1, January/February 2022, 190–201. <doi.org/10.1093/beheco/arab125>

22 Gero, S, Gordon, J & Whitehead, H. (2013). 'Calves as social hubs: dynamics of the social network within sperm whale units.' *Proceedings. Biological Sciences*, 280(1763), 20131113. <doi.org/10.1098/rspb.2013.1113>

23 Reddy, RB, Echelbarger, M, Toomajian, N, Hammond, T & Wellman, HM. (2023). 'Do children help dogs spontaneously?' *Human–Animal Interactions* (2023). <doi.org/10.1079/hai.2023.0001>

24 Goldenberg, SZ & Wittemyer, G. (2020). 'Elephant behavior toward the dead: A review and insights from field observations.' *Primates*, 61(1), 119–128.

25 Rutherford, L & Murray, LE. (2021). 'Personality and behavioral changes in Asian elephants (*Elephas maximus*) following the death of herd members.' *Integrative Zoology*, 16(2), 170–188.

26 Goldsborough, Z, Van Leeuwen, EJ, Kolff, KW, de Waal, FB & Webb, CE. (2020). 'Do chimpanzees (Pan troglodytes) console a bereaved mother?' *Primates*, 61, 93–102.

27 Reggente, MA, Alves, F, Nicolau, C, Freitas, L, Cagnazzi et al. (2016). 'Nurturant behavior toward dead conspecifics in free-ranging mammals: New records for odontocetes and a general review.' *Journal of Mammalogy*, 97(5), 1428–1434.

28 Rogers, T. (2016, August 16). 'Whale of a problem: Why do humpback whales protect other species from attack?' <theconversation.com/whale-of-a-problem-why-do-humpback-whales-protect-other-species-from-attack-63599>

Chapter 2 Why do we have sex?

1 Chromosomes Fact Sheet. (2020, August 15). <www.genome.gov/about-genomics/fact-sheets/Chromosomes-Fact-Sheet>

2 Ramesh, MA, Malik, SB & Logsdon, JM Jr. (2005). 'A phylogenomic inventory of meiotic genes; Evidence for sex in *Giardia* and an early eukaryotic origin of meiosis.' *Curr Biol.* 2005 Jan 26;15(2):185–91. <doi:10.1016/j.cub.2005.01.003>. PMID: 15668177.

3 Parfrey, LW & Lahr, DJ. (2013). 'Multicellularity arose several times in the evolution of eukaryotes (response to doi 10.1002/bies.201100187).' *Bioessays*, 35(4), 339–347.

4 Gibson, TM, Shih, PM, Cumming, VM, Fischer, WW, Crockford, PW et al. (2018). 'Precise age of *Bangiomorpha pubescens* dates the origin of eukaryotic photosynthesis.' *Geology*, 46(2), 135–138.

5 Krause, AJ et al. (2022). 'Extreme variability in atmospheric oxygen levels in the late Precambrian.' *Sci. Adv.* 8,eabm8191 (2022). <doi:10.1126/sciadv.abm8191>

6 Gold, DA, Grabenstatter, J, De Mendoza, A, Riesgo, A, Ruiz-Trillo, I & Summons, RE. (2016). 'Sterol and genomic analyses validate the sponge biomarker hypothesis.' *Proceedings of the National Academy of Sciences*, 113(10), 2684–2689.

7 Otto, SP. (2021).'Selective interference and the evolution of sex.' *Journal of Heredity*, 112(1), January 2021, 9–18, <doi.org/10.1093/jhered/esaa026>

8 Ryder, OA, Thomas, S, Judson, J M, Romanov, MN, Dandekar, S, Papp, JC, ... & Chemnick, LG. (2021). 'Facultative parthenogenesis in California condors.' Journal of Heredity, 112(7), 569–574.

9 Hörandl, E & Hadacek, F. (2020).'Oxygen, life forms, and the evolution of sexes in multicellular eukaryotes.' *Heredity* 125(1-2) (2020): 1–14.

Chapter 3 Why do we have males, females and other sexes?

1 Tong, SJW, & Ong, RS. (2020). 'Mating behavior, spawning, parental care, and embryonic development of some marine pseudocerotid flatworms (Platyhelminthes: Rhabditophora: Polycladida) in Singapore.' *Invertebrate Biology*, 139(2), e12293.

2 Peris, D et al. (2022). 'Large-scale fungal strain sequencing unravels the molecular diversity in mating loci maintained by long-term balancing selection.' *PLoS Genetics* 18.3 (2022): e1010097.

3 Bateman, AJ. (1948). 'Intra-sexual selection in *Drosophila*.' *Heredity* 2, 349–368 (1948).

4 Trivers, RL. (1972). 'Parental investment and sexual selection' in B Campbell, ed., *Sexual Selection and the Descent of Man 1871–1971*. London: Heinemann.

5 Bray, OE, Kennelly, JJ & Guarino, JL. (1975). 'Fertility of eggs produced on territories of vasectomized red-winged blackbirds.' *The Wilson Bulletin*, 87(2), 187–195. <www.jstor.org/stable/4160617>

6 Smith, SM. (1988). 'Extra-pair copulations in black-capped chickadees: The role of the female. Behaviour, 107(1-2), 15–23. <doi.org/10.1163/156853988x00160>

7 Gerlach, NM, McGlothlin, JW, Parker, PG & Ketterson, ED. (2012). 'Reinterpreting Bateman gradients: Multiple mating and selection in both sexes of a songbird species.' *Behavioral Ecology*, 23(5), September-October 2012, 1078–1088, <doi.org/10.1093/beheco/ars077>

8 Ibid.

9 Yasui, Y & Yamamoto, Y. (2021). 'An empirical test of bet-hedging polyandry hypothesis in the field cricket Gryllus bimaculatus.' *J Ethol* 39, 329–342 (2021). <doi.org/10.1007/s10164-021-00707-0>

10 Tang-Martinez, Z & Brandt Ryder, T. (2005). 'The problem with paradigms: Bateman's worldview as a case study.' *Integrative and Comparative Biology*, 45(5), November 2005, 821–830, <doi.org/10.1093/icb/45.5.821>. Tang-Martínez, Z. (2016). 'Rethinking Bateman's principles: Challenging persistent myths of sexually reluctant females and promiscuous males.' *J Sex Res*. 2016 May-Jun; 53(4-5):532–59. <doi:10.1080/00224499.2016.1150938>. Epub 2016 Apr 13. PMID: 27074147.

11 Gowaty, PA, Kim, YK & Anderson, WW. (2012). 'No evidence of sexual selection in a repetition of Bateman's classic study of *Drosophila melanogaster*.' *Proc Natl Acad Sci U S A*. 2012 Jul 17; 109(29):11740–5. <doi:10.1073/pnas.1207851109>. Epub 2012 Jun 11. PMID: 22689966; PMCID: PMC3406809.

12 Janicke, T, Häderer, IK, Lajeunesse, MJ & Anthes, N. (2016). 'Darwinian sex roles confirmed across the animal kingdom.' *Sci Adv*. 2016 Feb

12;2(2):e1500983. <doi:10.1126/sciadv.1500983>. PMID: 26933680; PMCID: PMC4758741.

13 Lehtonen, J. (2022). 'Bateman gradients from first principles.' *Nature Communications* 13.1 (2022): 3591.

14 Tang-Martinez, Z & Brandt Ryder, T. (2005). 'The problem with paradigms: Bateman's worldview as a case study.' *Integrative and Comparative Biology*, 45(5), November 2005, 821–830.

15 Lidborg, LH, Cross, CP & Boothroyd, LG. (2022). 'A meta-analysis of the association between male dimorphism and fitness outcomes in humans.' *Elife*. 2022 Feb 18;11:e65031. <doi: 10.7554/eLife.65031>. PMID: 35179485; PMCID: PMC9106334.

16 Pawłowski, B & Żelaźniewicz, A. (2021), 'The evolution of perennially enlarged breasts in women: A critical review and a novel hypothesis.' *Biological Reviews* 96.6 (2021): 2794–2809. Lassek, WD & Gaulin, SJC. (2022). 'Substantial but misunderstood human sexual dimorphism results mainly from sexual selection on males and natural selection on females.' *Frontiers in Psychology* 13 (2022): 859931.

17 Parker, GA., Baker, RR., & Smith, VGF. (1972). 'The origin and evolution of gamete dimorphism and the male-female phenomenon.' *Journal of Theoretical Biology*, *36*(3), 529–553. <doi.org/10.1016/0022-5193(72)90007-0>

18 Roughgarden, J. (2009). 'The Genial Gene: Deconstructing Darwinian Selfishness.' Google Books (p. 98). <www.google.com.au/books/edition/The_Genial_Gene/An8kDQAAQBAJ?hl=enandgbpv%3D1andpg%3DPA2andprintsec%3Dfrontcover&gbpv=1>

19 Roughgarden, J. (2009). 'The Genial Gene: Deconstructing Darwinian Selfishness.' Google Books (p. 99). <www.google.com.au/books/edition/The_Genial_Gene/An8kDQAAQBAJ?hl=enandgbpv=1andpg=PA2andprintsec=frontcover>.

20 Yasui, Y, & Hasegawa, E. (2022). 'The origination events of gametic sexual reproduction and anisogamy.' *Journal of Ethology*, *40*(3), 273–284.

21 *Springer Nature*. (2023). '2022 Evolutionary biology: A selection of 2022's highlighted research.' Kagawa-U.ac.jp. <www.ag.kagawa-u.ac.jp/wp-content/uploads/2023/04/Evolutionary-Biology-_-For-Researchers-_-Springer-Nature-2.html>

22 Wigby, S & Chapman, T. 'Sperm competition.' (2004). *Curr Biol*. 2004 Feb 3;14(3):R100-2. PMID: 14986632.

23 Hopkins, BR, and Perry, JC. (2022). 'The evolution of sex peptide: Sexual conflict, cooperation, and coevolution.' *Biol Rev Camb Philos Soc*. 2022 Aug;97(4):1426–1448. <doi:10.1111/brv.12849>. Epub 2022 Mar 6. PMID: 35249265; PMCID: PMC9256762.

24 Baruffaldi, L & Andrade, MCB. (2017). 'Neutral fitness outcomes contradict inferences of sexual "coercion" derived from male's damaging mating tactic in a widow spider.' *Sci Rep* 7, 17322 (2017). <doi.org/10.1038/s41598-017-17524-6>

25 Parker, GA., Baker, RR., & Smith, VGF. (1972). 'The origin and evolution of gamete dimorphism and the male-female phenomenon.' *Journal of Theoretical Biology*, *36*(3), 529–553. <doi.org/10.1016/0022-5193(72)90007-0>

26 Trivers, RL. (1972). 'Parental investment and sexual selection' in B Campbell, ed., *Sexual Selection and the Descent of Man 1871–1971*. London: Heinemann.

27 Fine, C. (2017). *Testosterone Rex: Myths of sex, science, and society*. <openlibrary.org/books/OL31443486M/Testosterone_Rex Chapter 2> Page 49.

28 Brown, GR, Laland, KN, & Mulder, MB. (2009). 'Bateman's principles and human sex roles.' *Trends in Ecology & Evolution*, 24(6), 297–304. <doi.org/10.1016/j.tree.2009.02.005>

29 Luoto, S & Correa Varella, MA. (2021). 'Pandemic leadership: Sex differences and their evolutionary–developmental origins.' *Frontiers in Psychology* 12 (2021): 633862.

30 <www.youtube.com/watch?app=desktopandv=0nTEfx44pws> (Video since deleted.)

31 *Species*. (2022, March 24). <nimpis.marinepests.gov.au/species/species/105>

32 Calder, WA. (1979). 'The kiwi and egg design: Evolution as a package deal.' *Bioscience* 29, 461–467.

33 Koçillari, L, Cattelan, S, Rasotto, MB, Seno, F, Maritan, A, & Pilastro, A. (2024). 'Tetrapod sperm length evolution in relation to body mass is shaped by multiple trade-offs.' *Nature Communications*, 15(1), 6160.

34 Whittington, CM, Griffith, OW, Qi, W, Thompson, MB & Wilson, AB. (2015). 'Seahorse brood pouch transcriptome reveals common genes associated with vertebrate pregnancy.' *Molecular Biology and Evolution*, 32(12), December 2015, 3114–3131.

35 Jukema, J & Piersma, T. (2006). 'Permanent female mimics in a lekking shorebird.' *Biology Letters*, 2.2 (2006): 161–164.

36 Feltman, R. (2015, November 16). 'A "supergene" turns these male birds into female impersonators or sneaky mate thieves – for life.' *Washington Post*. <www.washingtonpost.com/news/speaking-of-science/wp/2015/11/16/female-impersonator-or-mate-thief-for-these-birds-one-supergene-determines-sexual-strategy>

37 Galloway, R. (2017, June 8). 'How does a duck change its sex?' *BBC News*. <www.bbc.com/news/science-environment-40016817>

38 Lambert, MR, Tran, T, Kilian, A, Ezaz, T & Skelly, DK. (2019). 'Molecular evidence for sex reversal in wild populations of green frogs (*Rana clamitans*).' *Peer J*. 2019 Feb 8;7:e6449. <doi:10.7717/peerj.6449> PMID: 30775188; PMCID: PMC6369831.

39 Kean, Z. (2019, July 11). 'Stressing out and changing sex, researchers dive deep into the science of sex in bluehead wrasse.' ABC News.

<www.abc.net.au/news/science/2019-07-11/bluehead-wrasse-change-sex-epigenetic-stress/11291332>

40 Todd, EV, Ortega-Recalde, O, Liu, H, Lamm, MS, Rutherford, KM, Cross, H, ... & Gemmell, NJ. (2019). 'Stress, novel sex genes, and epigenetic reprogramming orchestrate socially controlled sex change.' *Science Advances*, *5*(7), eaaw7006.

41 Cameron was actually Head of School at the University of Tasmania while I was studying there – but we cannot judge her too harshly for it. The term intersex did not even appear in our major textbook! And she made me challenge many other assumptions.

42 James, S et al. (2019). 'Tracing the rise of malignant cell lines: Distribution, epidemiology and evolutionary interactions of two transmissible cancers in Tasmanian devils.' *Evolutionary Applications* 12.9 (2019): 1772–1780.

43 Marinov, GK. (2020). 'In Humans, Sex is Binary and Immutable.' *Academic Questions*, 33(2). <www.nas.org/academic-questions/33/2/in-humans-sex-is-binary-and-immutable>

44 Roughgarden, J. (2013). *Evolution's Rainbow: Diversity, gender, and sexuality in nature and people*. Berkeley: University Of California Press. <www.ucpress.edu/book/9780520280458/evolutions-rainbow> Cattet, M. (1988). 'Abnormal sexual differentiation in black bears (*Ursus americanus*) and brown bears (*Ursus arctos*).' *Journal of Mammalogy*, vol. 69, no. 4, 1988, 849–52. JSTOR <doi.org/10.2307/1381646>

45 Levy, T et al. (2020). 'Two homogametic genotypes – one crayfish: On the consequences of intersexuality.' *Iscience* 23.11 (2020).

46 McLaughlin, JF, Brock, KM, Gates, I, Pethkar, A, Piattoni, M, Rossi, A, & Lipshutz, SE. (2023). 'Multivariate models of animal sex: Breaking binaries leads to a better understanding of ecology and evolution.' *Integrative and Comparative Biology*, *63*(4), 891–906. <www.biorxiv.org/content/10.1101/2023.01.26.525769v1>

47 Ainsworth, C. (2015). 'Sex redefined.' *Nature* 518, 288–291 (2015). <doi.org/10.1038/518288a> King, DE. (2022). 'The inclusion of sex and gender beyond the binary in toxicology.' *Front Toxicol.* 2022 Jul 22;4:929219. <doi:10.3389/ftox.2022.929219>. PMID: 35936387; PMCID: PMC9355551.

48 Carpenter, M. (2017, March 10). 'Darlington Statement'. Intersex Human Rights Australia. <ihra.org.au/darlington-statement/>

49 Goymann, W, Brumm, H & Kappeler, PM. (2023). 'Biological sex is binary, even though there is a rainbow of sex roles: Denying biological sex is anthropocentric and promotes species chauvinism.' *Bioessays* 45.2 (2023): 2200173.

50 Fuentes, A. (2023, May 1). 'Here's Why Human Sex Is Not Binary.' *Scientific American*.<www.scientificamerican.com/article/heres-why-human-sex-is-not-binary/#:~:text=The%20bottom%20line%20is%20that>

Chapter 4 Why do we make love?

1 Soble, A. (2009). 'A history of erotic philosophy.' *Journal of Sex Research* 46.2-3 (2009): 104–120.

2 Massey, GJ. (1999). 'Medieval sociobiology: Thomas Aquinas's theory of sexual morality.' *Philosophical Topics* 27.1 (1999): 69–86. No. 1, 1999, 69–86. JSTOR, <www.jstor.org/stable/43154302>.

3 Ibid.

4 Marshall, S, & Miller, L. (n.d.). *You're Wrong About: Lesbian Seagulls with Lulu Miller* [Podcast]. You're Wrong About. <www.buzzsprout.com/1112270/13026519-lesbian-seagulls-with-lulu-miller?t=0>

5 Vines, G. (1999, August 7). 'Queer creatures.' *New Scientist*. <www.newscientist.com/article/mg16321985-000-queer-creatures>

6 Rodrigues, M. (2021, September 15). 'Same-Sex Sexual Behavior in Chimpanzees Challenge Our Gendered Biases About Evolution.' <www.prosocial.world/posts/same-sex-sexual-behavior-in-chimpanzees-challenge-our-gendered-biases-about-evolution>

7 Ungerfeld, R et al. (2013). 'Does heterosexual experience matter for bucks' homosexual mating behavior?' *Journal of Veterinary Behavior* 8.6 (2013): 471–474.

8 Monk, JD, Giglio, E, Kamath, A, Lambert, MR, & McDonough, CE. (2019). 'An alternative hypothesis for the evolution of same-sex sexual behaviour in animals.' *Nature ecology & evolution*, 3(12), 1622–1631.

9 Lerch, BA, & Servedio, MR. (2021). 'Same-sex sexual behaviour and selection for indiscriminate mating.' *Nature Ecology & Evolution*, 5(1), 135–141. <doi.org/10.1038/s41559-020-01331-w>

10 Brennan, PLR. (2022). 'Evolution and morphology of genitalia in female amniotes.' *Integrative and Comparative Biology*, 62(3), September 2022, 521–532.

11 Jannini, EA et al. (2012). 'Female orgasm(s): One, two, several.' *The Journal of Sexual Medicine* 9.4 (2012): 956–965.

12 Komisaruk, BR & Whipple, B. (2011). 'Non-genital orgasms.' *Sexual and Relationship Therapy*, 26:4, 356–372, <www.tandfonline.com/doi/full/10.1080/14681994.2011.649252?scroll=topandneedAccess=true>

13 O'Connell, HE et al. (1998). 'Anatomical relationship between urethra and clitoris.' *The Journal of Urology* 159.6 (1998): 1892–1897.

14 Ibid.

15 O'Connell, HE, Sanjeevan, KV & Hutson, JM. (2005). 'Anatomy of the clitoris.' *The Journal of Urology*, 174(4), 1189–1195.

16 Wise, NJ, Frangos, E & Komisaruk, BR. (2017). 'Brain activity unique to orgasm in women: An fMRI analysis.' *The Journal of Sexual Medicine*, 14(11), 1380–1391. <doi.org/10.1016/j.jsxm.2017.08.014>

17 Chadwick, SB, Francisco, M & van Anders, SM. (2019). 'When orgasms do not equal pleasure: Accounts of "bad" orgasm experiences during consensual sexual encounters.' *Arch Sex Behav* 48, 2435–2459 (2019). <doi.org/10.1007/s10508-019-01527-7>

18 Huynh, HK, Willemsen, AT & Holstege, G. (2013). 'Female orgasm but not male ejaculation activates the pituitary. A PET-neuro-imaging study.' *Neuroimage*, 76, 178–182.

19 Basanta, S, & Nuño de la Rosa, L. (2023). 'The female orgasm and the homology concept in evolutionary biology.' *Journal of Morphology*, *284*(1), e21544. <doi.org/10.1002/jmor.21544>

20 Basanta, S, & Nuño de la Rosa, L. (2023). The female orgasm and the homology concept in evolutionary biology. *Journal of Morphology*, *284*(1), e21544. </doi.org/10.1002/jmor.21544>

21 Pavličev, M, & Wagner, G. (2016). 'The evolutionary origin of female orgasm.' *Journal of Experimental Zoology Part B: Molecular and Developmental Evolution*, 326(6), 326–337.

22 Brennan, PLR. (2022). 'Evolution and morphology of genitalia in female amniotes.' *Integrative and Comparative Biology*, 62(3), September 2022, 521–532, <doi.org/10.1093/icb/icac115>

23 Pavličev, M et al. (2019). 'An experimental test of the ovulatory homolog model of female orgasm.' *Proceedings of the National Academy of Sciences* 116.41 (2019): 20267–20273.

24 Lewis, T. (2019, December 1). 'Can Rabbits Help Unravel the Mystery of Female Orgasm?' *Scientific American*. <www.scientificamerican.com/article/can-rabbits-help-unravel-the-mystery-of-female-orgasm/>

25 Mahar, EA, Mintz, LB & Akers, BM. (2020). 'Orgasm equality: Scientific findings and societal implications.' *Current Sexual Health Reports*, 12, 24–32.

26 Pavličev, & Wagner, G. (2016). 'The evolutionary origin of female orgasm.' *Journal of Experimental Zoology Part B: Molecular and Developmental Evolution* 326.6 (2016): 326–337.

27 Brennan, PLR, Cowart, R & Orbach, DN. (2022). 'Evidence of a functional clitoris in dolphins.' *Current Biology* 32.1 (2022): R24–R26.

28 Brennan, P. (2022, January 11), Functional morphology of the dolphin clitoris.' *Current Biology*. Jan. 10. 2022 (Vol. 32, Issue 1) (M. Mix, interviewer). <www.youtube.com/watch?v=MprSbmrkdpo>

29 Bagemihl, B. (1999). *Biological Exuberance: Animal homosexuality and natural diversity*. New York: Macmillan. Gaşpar, C, Ailincăi, LI, and Dodan, AX. (2022). 'Observations of sexual behaviors in goats (*Capra hircus*) raised on non-professional farms.' *Journal of Applied Life Sciences and Environment*, 191(3).

30 Rogers, & Gibbs, RA. (2014). 'Comparative primate genomics: Emerging patterns of genome content and dynamics.' *Nature Reviews Genetics* 15.5 (2014): 347–359.

31 Wrangham, RW. (1993). 'The evolution of sexuality in chimpanzees and bonobos.' *Human Nature*, 4, 47–79.

32 Genty, E, Neumann, C & Zuberbühler, K. (2015). 'Complex patterns of signalling to convey different social goals of sex in bonobos, *Pan paniscus*.' *Scientific Reports* 5.1 (2015): 16135.

33 Moscovice, LR, Surbeck, M, Fruth, B, Hohmann, G, Jaeggi, AV & Deschner, T. (2019). 'The cooperative sex: Sexual interactions among female bonobos are linked to increases in oxytocin, proximity and coalitions.' *Hormones and Behavior*, 116, 104581.

34 Barron, AB & Hare, B. (2020). 'Prosociality and a sociosexual hypothesis for the evolution of same-sex attraction in humans.' *Front. Psychol.* 2020 Jan 16;10:2955. <doi:10.3389/fpsyg.2019.02955>. PMID: 32010022; PMCID: PMC6976918.

35 Reis, HT et al. (2013). 'Ellen Berscheid, Elaine Hatfield, and the emergence of relationship science.' *Perspectives on Psychological Science* 8.5 (2013): 558–572.

36 McLennan, DA. (2008). 'The concept of co-option: Why evolution often looks miraculous.' *Evolution: Education and Outreach, 1*, 247–258.

37 Bode, A, & Kavanagh, PS. (2023). 'Romantic love and behavioral activation system sensitivity to a loved one.' *Behavioral Sciences*, 13(11), 921.

38 McLennan, DA. (2008). 'The concept of co-option: Why evolution often looks miraculous.' *Evolution: Education and Outreach* 1.3 (2008): 247–258.

39 Mitani, JC. (2009). 'Male chimpanzees form enduring and equitable social bonds.' *Animal Behaviour* 77.3 (2009): 633–640.

Chapter 5 Why do we get cancer?

1 Cancer Australia. (2022). Cancer in Australia Statistics. <www.canceraustralia.gov.au/impacted-cancer/what-cancer/cancer-australia-statistics>

2 Thomas, F, Madsen, T, Giraudeau, M, Misse, D, Hamede, R, Vincze, O, … & Ujvari, B. (2019). 'Transmissible cancer and the evolution of sex.' *PLOS Biology*, 17(6), e3000275. <doi.org/10.1371/journal.pbio.3000275>

3 Brüniche-Olsen, A, Jones, ME, Burridge, CP, Murchison, EP, Holland, BR, & Austin, JJ. (2018). 'Ancient DNA tracks the mainland extinction and island survival of the Tasmanian devil.' *Journal of Biogeography*, *45*(5), 963–976. <doi.org/10.1111/jbi.13214>

4 Pearse, A. M., & Swift, K. (2006). 'Transmission of devil facial-tumour disease.' *Nature*, *439*(7076), 549–549. <www.nature.com/articles/439549a>

5 Murchison, E. P. (2016). Transmissible tumours under the sea. *Nature*, *534*(7609), 628–629. <www.nature.com/articles/nature18455>

6 Cunningham, CX, Comte, S, McCallum, H, Hamilton, DG, Hamede, R, Storfer, A, ... & Jones, ME. (2021). 'Quantifying 25 years of disease-caused declines in Tasmanian devil populations: Host density drives spatial pathogen spread.' *Ecology Letters*, *24*(5), 958–969. <onlinelibrary.wiley.com/doi/abs/10.1111/ele.13703>

7 Jessy, T. (2001). 'Immunity over inability: The spontaneous regression of cancer.' *J Nat Sci Biol Med.* 2011 Jan;2(1):43–9. <doi:10.4103/0976-9668.82318>. PMID: 22470233; PMCID: PMC3312698. Fauvet, J et al. (1960). 'Cures, regressions and spontaneous remissions of cancer.' *La Revue du Praticien.* 10 (1960): 2349–84.

8 Pye, R, Hamede, R, Siddle, HV, Caldwell, A, Knowles, GW, Swift, K, ... & Woods, GM. (2016). 'Demonstration of immune responses against devil facial tumour disease in wild Tasmanian devils.' *Biology letters*, 12(10), 20160553. <royalsocietypublishing.org/doi/full/10.1098/rsbl.2016.0553>

9 Vendramin, R, Litchfield, K & Swanton, C. (2021). 'Cancer evolution: Darwin and beyond.' *EMBO J.* 2021 Sep 15;40(18):e108389. <doi:10.15252/embj.2021108389>. Epub 2021 Aug 30. PMID: 34459009; PMCID: PMC8441388.

10 Murchison, EP, Wedge, DC, Alexandrov, LB, Fu, B, Martincorena, I, Ning, Z, ... & Stratton, MR. (2014). 'Transmissible dog cancer genome reveals the origin and history of an ancient cell lineage.' *Science*, *343*(6169), 437440. <10.1126/science.1247167>

11 Odes, EJ, Randolph-Quinney, PS, Steyn, M, Throckmorton, Z, Smilg, JS, Zipfel, B, ... & Berger, LR. (2016). 'Earliest hominin cancer: 1.7-million-year-old osteosarcoma from Swartkrans Cave, South Africa.' *South African Journal of Science*, 112(7-8), 1–5. <dx.doi.org/10.17159/sajs.2016/20150471>

12 Odes, EJ, & Randolph-Quinney, P. (2016, September 13). 'Fossil evidence reveals that cancer in humans goes back 1.7 million years.' *The Conversation.* <theconversation.com/fossil-evidence-reveals-that-cancer-in-humans-goes-back-1-7-million-years-63430>

13 Haridy, Y, Witzmann, F, Asbach, P, Schoch, RR, Fröbisch, N, & Rothschild, BM. (2019). 'Triassic cancer – osteosarcoma in a 240-million-year-old stem-turtle.' *JAMA oncology*, *5*(3), 425–426.

14 Ekhtiari, S, Chiba, K, Popovic, S, Crowther, R, Wohl, G, Wong, AKO, ... & Evans, DC. (2020). 'First case of osteosarcoma in a dinosaur: A multimodal diagnosis.' *The Lancet Oncology*, *21*(8), 1021–1022. <www.researchgate.net/publication/343411126_First_case_of_osteosarcoma_in_a_dinosaur_a_multimodal_diagnosis>

15 Aktipis, CA, Boddy, AM, Jansen, G, Hibner, U, Hochberg, ME, Maley, CC, & Wilkinson, GS. (2015). 'Cancer across the tree of life: Cooperation and cheating in multicellularity.' *Philosophical transactions of the Royal Society of London. Series B, Biological sciences*, *370*(1673), 20140219. <doi.org/10.1098/rstb.2014.0219>

16 For the detail of Aktipis's criteria for cancer see her paper: Aktipis, CA, Boddy, AM, Jansen, G, Hibner, U, Hochberg, ME., Maley, CC, & Wilkinson, GS. (2015). 'Cancer across the tree of life: cooperation and cheating in multicellularity.' *Philosophical transactions of the Royal Society of London.* Series B, Biological sciences, 370(1673), 20140219.

<doi.org/10.1098/rstb.2014.0219> and her book: Aktipis, A. (2021). *The Cheating Cell: How evolution helps us understand and treat cancer*. Princeton: Princeton University Press.

17 Kwiatkowski, F, Arbre, M, Bidet, Y, Laquet, C, Uhrhammer, N, & Bignon, YJ. (2015). 'BRCA mutations increase fertility in families at hereditary breast/ovarian cancer risk.' *PLoS One*, 10(6), e0127363. <10.1371/journal.pone.0127363>

18 Nunney, L. (2018). 'Size matters: height, cell number and a person's risk of cancer.' *Proceedings of the Royal Society* B, 285(1889), 20181743.

19 Caulin, AF, & Maley, CC. (2011). 'Peto's Paradox: Evolution's prescription for cancer prevention.' *Trends in ecology & evolution*, *26*(4), 175–182. <doi.org/10.1016/j.tree.2011.01.002>

20 Olivier, M, Hollstein, M, & Hainaut, P. (2010). 'TP53 mutations in human cancers: Origins, consequences, and clinical use.' *Cold Spring Harbor perspectives in biology*, *2*(1), a001008. <doi.org/10.1101/cshperspect.a001008>

21 'The tumour suppressor.' (2022, March 2). Nature, p. 53. <www.nature.com/collections/faifcfdeic>

22 Schneider, K, Zelley, K, Nichols, KE, & Garber, J. (2019, November 21). 'Li-Fraumeni Syndrome.' Nih. <www.ncbi.nlm.nih.gov/books/NBK1311/>

23 <www.cancer.net/cancer-types/li-fraumeni-syndrome>

24 Preston, AJ, Rogers, A, Sharp, M, Mitchell, G, Toruno, C et al. (2023). 'Elephant TP53-RETROGENE 9 induces transcription-independent apoptosis at the mitochondria.' *Cell Death Discovery*, 9(1), 66.

25 Preston, AJ, Rogers, A, Sharp, M, Mitchell, G, Toruno, C, Barney, BB., ... & Abegglen, LM. (2023). 'Elephant TP53-RETROGENE 9 induces transcription-independent apoptosis at the mitochondria.' *Cell Death Discovery*, *9*(1), 66. <doi.org/10.1038/s41420-023-01348-7>

26 Rojas, LA, Sethna, Z, Soares, KC, Olcese, C, Pang, N, Patterson, E, ... & Balachandran, VP. (2023). 'Personalized RNA neoantigen vaccines stimulate T cells in pancreatic cancer.' *Nature*, *618*(7963), 144–150. <doi.org/10.1038/s41586-023-06063-y>

27 Compton, ZT, Harris, V, Mellon, W, Rupp, S, Mallo, D et al. (2023). 'Cancer prevalence across vertebrates.' *Research Square*. <www.ncbi.nlm.nih.gov/pmc/articles/PMC10350200/>

28 Cancer Australia. (2019, December 18). 'Prostate cancer in Australia statistics.' <www.canceraustralia.gov.au/cancer-types/prostate-cancer/statistics>

29 Zhang, J, Cunningham, J, Brown, J, & Gatenby, R. (2022). 'Evolution-based mathematical models significantly prolong response to abiraterone in metastatic castrate-resistant prostate cancer and identify strategies to further improve outcomes.' *Elife*, *11*, e76284. <doi.org/10.7554/eLife.76284>

30 Gatenby, RA., Zhang, J, & Brown, JS. (2019). 'First strike–second strike strategies in metastatic cancer: Lessons from the evolutionary dynamics of extinction.' *Cancer research*, *79*(13), 3174–3177. <aacrjournals.org/cancerres/article/79/13/3174/634088/First-Strike-Second-Strike-Strategies-in>
31 Gatenby, RA, Zhang, J, & Brown, JS. (2019). 'First strike–second strike strategies in metastatic cancer: Lessons from the evolutionary dynamics of extinction.' *Cancer research*, 79(13), 3174–3177.
32 Artzy-Randrup, Y, Epstein, T, Brown, JS, Costa, RL, Czerniecki, BJ, & Gatenby, RA. (2021). 'Novel evolutionary dynamics of small populations in breast cancer adjuvant and neoadjuvant therapy'. *NPJ Breast Cancer*, *7*(1), 26. <www.nature.com/articles/s41523-021-00230-y>
33 Ibid.
34 Heikamp, EB & Pui, CH. (2018). 'Next-generation evaluation and treatment of pediatric acute lymphoblastic leukemia'. *The Journal of Pediatrics*, 203, 14–24.e2. <10.1016/j.jpeds.2018.07.039>
35 *New Clinical Trial for Personalised Adaptive Treatment of Prostate Cancer Enrolls First Participants*. (n.d.). <anzup.org.au/wp-content/uploads/2023/06/ANZadapt-first-patients-MR-FINAL.pdf>
36 Margres, MJ, Ruiz-Aravena, M, Hamede, R, Chawla, K, Patton, AH et al. (2020). 'Spontaneous tumor regression in Tasmanian devils associated with RASL11A activation.' *Genetics*, 215(4), 1 August 2020, 1143–1152.<doi.org/10.1534/genetics.120.303428> Louro, R, Nakaya, HI, Paquola, AC, Martins, EA, da Silva, AM et al. (2004). 'RASL11A, member of a novel small monomeric GTPase gene family, is down-regulated in prostate tumors.' *Biochem Biophys Res Commun*. 2004 Apr 9;316(3):618–27. <doi: 10.1016/j.bbrc.2004.02.091>. PMID: 1503344.

Chapter 6 Why do we age?

1 Kyriazis, M. (2020). 'Ageing as "time-related dysfunction": A perspective.' *Front Med* (Lausanne). 2020 Jul 27;7:371. <doi:10.3389/fmed.2020.00371>. PMID: 32850891; PMCID: PMC7397818.
2 Ebeling, M, Rau, R, Malmström, H, Ahlbom, A & Modig, K. 'The rate by which mortality increases with age is the same for those who experienced chronic disease as for the general population.' *Age and Ageing*. 2021 Sep 11;50(5):1633–1640. <doi:10.1093/ageing/afab085>. PMID: 34038514; PMCID: PMC8437060.
3 López-Otín, C, Blasco, MA, Partridge, L, Serrano, M, & Kroemer, G. (2013). 'The hallmarks of aging.' *Cell*, 153(6), 1194–1217. <doi.org/10.1016/j.cell.2013.05.039>
4 Ferrucci, L & Fabbri, E. (2018). 'Inflammageing: Chronic inflammation in ageing, cardiovascular disease, and frailty'. *Nat Rev Cardiol*. 2018 Sep;15(9):505–522. <doi:10.1038/s41569-018-0064-2>. PMID: 30065258; PMCID: PMC6146930.

5 López-Otín, C, Blasco, MA, Partridge, L, Serrano, & Kroemer, G.(2023). 'Hallmarks of ageing: An expanding universe.' *Cell* 186(2), 19 January 2023, 243–278.
6 Sinclair, D, Laplante, MD, & Delphia, C. (2019). *Lifespan: Why we age – and why we don't have to*. New York: Atria Books.
7 Lemoine, M. (2021). 'The evolution of the hallmarks of aging.' *Frontiers in Genetics*, 12, 693071.
8 Animals like sea stars and snails that are not symmetrical evolved asymmetry from animals that were symmetrical with a head.
9 Kowalczyk, A, Partha, R, Clark, NL & Chikina, M. (2020). 'Pan-mammalian analysis of molecular constraints underlying extended life span.' *eLife* 9:e51089. <doi.org/10.7554/eLife.51089>
10 Lee, PC, Fishlock, V, Webber, CE & Moss, CJ. (2016). 'The reproductive advantages of a long life: Longevity and senescence in wild female African elephants.' *Behav Ecol Sociobiol*. 2016;70:337–345. <doi:10.1007/s00265-015-2051-5>. Epub 2016 Jan 19. PMID: 26900212; PMCID: PMC4748003.
11 Medawar, P. (1951). 'Medawar 1952 Unsolved Problem.' *Internet Archive*. <archive.org/details/medawar-1952-unsolved-problem/page/13/mode/2up>
12 de Magalhães, JP. (2024). 'The longevity bottleneck hypothesis: Could dinosaurs have shaped ageing in present-day mammals?' *BioEssays* 46.1 (2024): 2300098.
13 Chen, H & Maklakov, AA. (2012). 'Longer life span evolves under high rates of condition-dependent mortality.' Supplemental Information – *Current Biology* 22.22 (2012): 2140–2143.
14 Ibid.
15 Nussey, DH et al. (2013). 'Senescence in natural populations of animals: Widespread evidence and its implications for bio-gerontology.' *Ageing Research Reviews* 12.1 (2013): 214–225.
16 Bonduriansky, R, & Brassil, CE. (2002). 'Rapid and costly ageing in wild male flies.' *Nature*, 420(6914), 377–377.
17 Smith, D. (2013, June 17). 'Smith: World's oldest wild bear showing frailty at 39, but still roaming northern Minnesota.'
18 Ibid.
19 Williams, GC. (1957). 'Pleiotropy, natural selection, and the evolution of senescence.' *Evolution*, 11(4), 398–411. <doi.org/10.1111/j.1558-5646.1957.tb02911.x>
20 He, Q et al. (2014). 'Shorter men live longer: Association of height with longevity and FOXO3 genotype in American men of Japanese ancestry.' *PLoS One* 9.5 (2014): e94385. Chmielewski, PP. (2023). 'The association between body height and longevity: Evidence from a national population sample.' *Folia Morphologica* (2023).
21 Day, OF, Elks, C, Murray, A et al. (2015). 'Puberty timing associated with diabetes, cardiovascular disease and also diverse health outcomes

in men and women: The UK Biobank study.' *Sci Rep* 5, 11208 (2015). <doi.org/10.1038/srep11208>

22 Li, Stacy, Vazquez, JM & Sudmant, PH. (2023). 'The evolution of ageing and life span.' *Trends in Genetics* (2023).

23 Kirkwood, TB. (1977). 'Evolution of ageing.' *Nature*, 270(5635), 301–304.

24 Goldsmith, TC. (2008). 'Ageing, evolvability, and the individual benefit requirement; medical implications of ageing theory controversies.' *Journal of Theoretical Biology* 252.4 (2008): 764–768.

25 Wikenros C, Gicquel, M, Zimmermann, B, Flagstad, Ø & Åkesson, M. (2021). 'Age at first reproduction in wolves: Different patterns of density dependence for females and males.' *Proc. R. Soc.* B.28820210207210207 <doi.org/10.1098/rspb.2021.0207>

26 Sidorovich, VE, Stolyarov, VP, Vorobei, NN, Ivanova, NV & Jędrzejewska, B. (2007). 'Litter size, sex ratio, and age structure of gray wolves, *Canis lupus*, in relation to population fluctuations in northern Belarus.' *Canadian Journal of Zoology*. 85(2): 295–300. <doi.org/10.1139/Z07-001>

27 Zhu, P, Liu, W, Zhang, X, Li, M, Liu, G et al. (2023). 'Correlated evolution of social organization and life span in mammals.' *Nature Communications*, 14(1), 372.

28 de Beer, J, Bardoutsos, A & Janssen, F. (2017). 'Maximum human life span may increase to 125 years.' *Nature*, 546(7660), E16-E17.

29 Sanghvi, K, Vega-Trejo, R, Nakagawa, S, Gascoigne, SJ, Johnson, SL, Salguero-Gómez, R, ... & Sepil, I. (2024). 'Meta-analysis shows no consistent evidence for senescence in ejaculate traits across animals.' *Nature Communications*, 15(1), 558.

30 Ruggeri, A. (2018). 'Do we really live longer than our ancestors?' BBC. <www.bbc.com/future/article/20181002-how-long-did-ancient-people-live-life-span-versus-longevity>

31 Carrieri, MP & Serraino, D. (2005). 'Longevity of popes and artists between the 13th and the 19th century.' *International Journal of Epidemiology*, 34(6), December 2005, 1435–1436, <doi.org/10.1093/ije/dyi211>

32 McCauley, B. (2018). 'Life expectancy in hunter-gatherers.' *Encyclopedia of Evolutionary Psychological Science*, 1–3. <doi.org/10.1007/978-3-319-16999-6_2352-1>

33 Hublin, JJ, Ben-Ncer, A, Bailey, SE, Freidline, SE, Neubauer, S et al. (2017). 'New fossils from Jebel Irhoud, Morocco and the pan-African origin of *Homo sapiens*.' *Nature*, 546(7657), 289–292.

34 Caspari, R & Lee, S-H. (2006). 'Is human longevity a consequence of cultural change or modern biology?' *American Journal of Physical Anthropology: The Official Publication of the American Association of Physical Anthropologists* 129.4 (2006): 512–517.

35 Caspari, R. (2012, December 1). 'The Evolution of Grandparents.'

Scientific American. <www.scientificamerican.com/article/the-evolution-of-grandparents-2012-12-07/>
36 Williams, GC. (1957). 'Pleiotropy, natural selection, and the evolution of senescence.' *Evolution*, 11(4), 398–411. <doi.org/10.1111/j.1558-5646.1957.tb02911.x>
37 Nattrass, S, Croft, DP, Ellis, S, Cant, MA, Weiss, MN et al. (2019). 'Postreproductive killer whale grandmothers improve the survival of their grandoffspring.' *Proceedings of the National Academy of Sciences*, 116(52), 26669–26673.
38 Hawkes, K, O'Connell, JF & Blurton Jones, NG. (1997). 'Hadza women's time allocation, offspring provisioning, and the evolution of long postmenopausal life spans.' *Current Anthropology*, 38(4), 551–577. <doi.org/10.1086/204646>
39 Caspari, R. (2012, December 1). 'The Evolution of Grandparents.' Scientific American. <www.scientificamerican.com/article/the-evolution-of-grandparents-2012-12-07/>
40 Cant, M. (2023). 'Menopause in chimpanzees.' *Science*, 382(6669), 368–369.
41 Sanghvi, K, Vega-Trejo, R, Nakagawa, S, Gascoigne, SJ, Johnson, SL et al. (2024). 'Meta-analysis shows no consistent evidence for senescence in ejaculate traits across animals.' *Nature Communications*, 15(1), 558.
42 Ibid.
43 Hooper, PL, Gurven, M, Winking, J & Kaplan, HS. (2015). 'Inclusive fitness and differential productivity across the life course determine intergenerational transfers in a small-scale human society.' *Proc. R. Soc.* B.28220142808 <doi.org/10.1098/rspb.2014.2808>
44 Chapman, S, Danielsbacka, M, Tanskanen, AO, Lahdenperä, M, Pettay, J & Lummaa, V. (2023). 'Grandparental co-residence and grandchild survival: The role of resource competition in a pre-industrial population.' *Behavioral Ecology*, 34(3), 446–456.
45 McComb, K, Shannon, G, Durant, SM, Sayialel, K, Slotow, R et al. (2011). 'Leadership in elephants: The adaptive value of age.' *Proceedings. Biological Sciences*, 278(1722), 3270–3276. <doi.org/10.1098/rspb.2011.0168>
46 Allen, CR, Brent, LJ, Motsentwa, T, Weiss, MN & Croft, DP. (2020). 'Importance of old bulls: Leaders and followers in collective movements of all-male groups in African savannah elephants (*Loxodonta africana*).' *Scientific Reports*, 10(1), 13996.
47 Cassidy, KA, MacNulty, DR, Stahler, DR, Smith, DW & Mech, LD. (2015). 'Group composition effects on aggressive interpack interactions of gray wolves in Yellowstone National Park.' *Behavioral Ecology*, 26(5), 1352–1360.
48 Davison, R & Gurven, M. (2022). 'The importance of elders: Extending Hamilton's force of selection to include intergenerational transfers.' *Proceedings of the National Academy of Sciences*, 119(28), e2200073119.

Chapter 7 Why do we drink?

1 Although how that is developed from childhood is still debated: LoBue, V, & Adolph, KE. (2019). 'Fear in infancy: Lessons from snakes, spiders, heights, and strangers.' *Developmental Psychology*, *55*(9), 1889–1907. < doi.org/10.1037/dev0000675>

2 Koob, GF & Volkow, ND. (2016). 'Neurobiology of addiction: A neurocircuitry analysis.' *The Lancet Psychiatry* 3.8 (2016): 760–773.

3 <nida.nih.gov/publications/research-reports/marijuana/how-does-marijuana-produce-its-effects>

4 Roque Bravo, R, Faria, AC, Brito-da-Costa, AM, Carmo, H, Mladěnka, P, Dias da Silva, D, & Remião, F. On Behalf Of The Oemonom Researchers. (2022). 'Cocaine: An updated overview on chemistry, detection, biokinetics, and pharmacotoxicological aspects including abuse pattern.' *Toxins*, 14(4), 278. <doi.org/10.3390/toxins14040278>

5 Mitchell, JM et al. (2012). 'Alcohol consumption induces endogenous opioid release in the human orbitofrontal cortex and nucleus accumbens.' *Science Translational Medicine* 4.116 (2012): 116ra6-116ra6.

6 No level of alcohol consumption is safe for our health. (n.d.). <www.who.int/europe/news/item/04-01-2023-no-level-of-alcohol-consumption-is-safe-for-our-health>

7 Wigger, GW et al. (2022). 'The impact of alcohol use disorder on tuberculosis: A review of the epidemiology and potential immunologic mechanisms.' *Frontiers in Immunology* 13 (2022): 864817.

8 Heart Foundation. (2021). Alcohol & Heart Health. <assets.contentstack.io/v3/assets/blt8a393bb3b76c0ede/blt0dce80111f132aef/65ea8e30aac51def878fa33a/210311_Evidence-paper_Summary_ALCOHOL.pdf.>

9 Clites, BL, Hofmann, HA, & Pierce, JT. (2023). 'The promise of an evolutionary perspective of alcohol consumption.' *Neuroscience Insights*, 18, 26331055231163589.

10 Dominy, NJ. (2004). "Fruits, fingers, and fermentation: The sensory cues available to foraging primates.' *Integrative and Comparative Biology*, 44(4), 295–303.

11 Dudley, R. (2002). 'Fermenting fruit and the historical ecology of ethanol ingestion: Is alcoholism in modern humans an evolutionary hangover?' *Addiction*, 97(4), 381–388.

12 Dudley, R. (2000). 'Evolutionary origins of human alcoholism in primate frugivory.' *The Quarterly Review of Biology*, 75(1), 3–15.

13 Milton, K. (2004). 'Ferment in the family tree: Does a frugivorous dietary heritage influence contemporary patterns of human ethanol use?' *Integrative and Comparative Biology*, 44(4), 304–314.

14 Amato, KR et al. (2021). 'Fermented food consumption in wild nonhuman primates and its ecological drivers.' *American Journal of Physical Anthropology* 175.3 (2021): 513–530.

15 Den Boer, W, Campione, NE., & Kear, BP. (2019). 'Climbing adaptations, locomotory disparity and ecological convergence in ancient stem "kangaroos".' *Royal Society Open Science*, *6*(2), 181617. <royalsocietypublishing.org/doi/full/10.1098/rsos.181617>

16 Böhme, M et al. (2019). 'A new Miocene ape and locomotion in the ancestor of great apes and humans.' *Nature* 575.7783 (2019): 489–493.

17 Retallack, G. J. (2023). Ecological polarities of African Miocene apes. *Evolving Earth*, *1*, 100005. <www.sciencedirect.com/science/article/pii/S2950117223000055>

18 Carrigan, MA et al. (2014). 'Hominids adapted to metabolize ethanol long before human-directed fermentation.' *Proc Natl Acad Sci USA* 112, 458–463.

19 Ibid.

20 Morris, S, Humphreys, D & Reynolds, D. (2006). 'Myth, marula, and elephant: An assessment of voluntary ethanol intoxication of the African elephant (*Loxodonta africana*) following feeding on the fruit of the marula tree (*Sclerocarya birrea*).' *Physiol Biochem Zool.* 2006 Mar-Apr;79(2):363–9. <doi:10.1086/499983>. Epub 2006 Feb 6. PMID: 16555195.

21 Janiak, MC, Pinto, SL, Duytschaever, G, Carrigan, MA & Melin, AD. (2020). 'Genetic evidence of widespread variation in ethanol metabolism among mammals: Revisiting the "myth" of natural intoxication.' *Biol. Lett.*162020007020200070 <doi.org/10.1098/rsbl.2020.0070>

22 Liu, L et al. (2018). 'Fermented beverage and food storage in 13,000 y-old stone mortars at Raqefet Cave, Israel: Investigating Natufian ritual feasting.' *Journal of Archaeological Science*: Reports 21 (2018): 783–793.

23 Wang, J, Jiang, L & Sun, H. (2021). 'Early evidence for beer drinking in a 9000-year-old platform mound in southern China.' *PLoS One* 16.8 (2021): e0255833.

24 Longrich, NR. (2021, July 16). 'When did humans start experimenting with alcohol and drugs?' *The Conversation*. <theconversation.com/when-did-humans-start-experimenting-with-alcohol-and-drugs-161556>

25 Varela, C, Alperstein, L, Sundstrom, J, Solomon, M, Brady, M, Borneman, A, & Jiranek, V. (2023). 'A special drop: Characterising yeast isolates associated with fermented beverages produced by Australia's indigenous peoples.' *Food Microbiology*, 112, 104216. <www.sciencedirect.com/science/article/abs/pii/S0740002023000035>

26 Machin, AJ, & Dunbar, RI. (2011). 'The brain opioid theory of social attachment: A review of the evidence.' *Behaviour*, 148(9-10), 985–1025.

27 Dunbar, RIM, Launay, J, Wlodarski, R, Robertson, C, Pearce, E, Carney, J et al. (2017). 'Functional benefits of (modest) alcohol consumption.' *Adapt Human Behav Physiol.* 2017;3(2):118–133. <doi:10.1007/s40750-016-0058-4>. Epub 2016 Dec 28. PMID: 32104646; PMCID: PMC7010365.

28 Young, E (2017, January 24). 'Iceland knows how to stop teen substance abuse – but the rest of the world isn't listening.' Triple J. <www.abc.net.au/triplej/programs/hack/iceland-teen-substance-abuse/8208214>

29 Cowan, R, & correspondent, I. (2001, May 17). 'Dancers fall in line for damnation, says Paisley.' *The Guardian*. <www.theguardian.com/uk/2001/may/18/ianpaisley> Free Presbyterian Church of Scotland Website Closed for the Sabbath – Free Presbyterian Church of Scotland. (2024). <www.fpchurch.org.uk/about-us/frequently-asked-questions/>

30 Li, H, Gu, S, Han, Y, Xu, Z, Pakstis, AJ, Jin, L et al. (2011). 'Diversification of the ADH1B gene during expansion of modern humans.' *Ann Hum Genet*. 2011 Jul;75(4):497–507. <doi: 10.1111/j.1469-1809.2011.00651>. 2011.00651.x. Epub 2011 May 18. PMID: 21592108; PMCID: PMC3722864.

31 Enoch, M-A & Albaugh, BJ. (2017). 'Genetic and environmental risk factors for alcohol use disorders in American Indians and Alaskan Natives.' *The American Journal on Addictions* 26.5 (2017): 461–468.

32 Zaridze, D et al. (2009). 'Alcohol and cause-specific mortality in Russia: A retrospective case–control study of 48 557 adult deaths.' *The Lancet* 373.9682 (2009): 2201–2214.

Chapter 8 Why do we sleep?

1 Long, JA et al. (2015). 'First shark from the Late Devonian (Frasnian) Gogo Formation, Western Australia sheds new light on the development of tessellated calcified cartilage.' *PLoS One* 10.5 (2015): e0126066.

2 Australian Government. (2019). Homepage | gbrmpa. <www2.gbrmpa.gov.au/>

3 Mukhametov, LM, Supin, AY & Polyakova, IG. (1077). 'Interhemispheric asymmetry of the electroencephalographic sleep patterns in dolphins.' Brain Res. 1977 Oct 14;134(3):581–4. <doi:10.1016/0006-8993(77)90835-6>. PMID: 902119.

4 Discovery. (2016, June 28). 'Great White Naps for First Time on Camera.' *YouTube*. <www.youtube.com/watch?v=B7ePdi1McMo>

5 Kelly, ML, Spreitzenbarth, S, Kerr, CC, Hemmi, JM, Lesku, JA et al. (2021). 'Behavioural sleep in two species of buccal pumping sharks (*Heterodontus portusjacksoni* and *Cephaloscyllium isabellum*).' *J Sleep Res*. 2021 Jun; 30(3):e13139. <doi:10.1111/jsr.13139>. Epub 2020 Jul 16. PMID: 32672393.

6 Kelly, ML et al. (2022). 'Energy conservation characterizes sleep in sharks.' *Biology Letters* 18.3 (2022): 20210259.

7 Everson, CA, Bergmann, BM & Rechtschaffen, A. (1989). 'Sleep deprivation in the rat: III. Total sleep deprivation.' *Sleep*. 1989 Feb;12(1):13–21. <doi:10.1093/sleep/12.1.13>. PMID: 2928622. Rechtschaffen, A &Bergmann, BM. (2002). 'Sleep deprivation in the rat: An update of the 1989 paper.' *Sleep*. 2002 Feb 1;25(1):18–24. <doi:10.1093/sleep/25.1.18>. PMID: 11833856.

8 Riddle, DL, Blumenthal, T, Meyer, BJ et al, eds. (1997). *C. elegans II*. 2nd edition. Cold Spring Harbor (NY): Cold Spring Harbor Laboratory Press; 1997. Section I, The Biological Model. Available from: <www.ncbi.nlm.nih.gov/books/NBK20086/>

9 De Robertis, EM & Tejeda-Muñoz, N. (2022). 'Evo-Devo of Urbilateria and its larval forms.' *Developmental Biology* 487 (2022): 10–20.

10 Kawano, T et al. (2023). 'ER proteostasis regulators cell-non-autonomously control sleep.' *Cell Reports* 42.3 (2023).

11 Cirelli, C. (2013). 'Sleep and synaptic changes.' *Curr Opin Neurobiol.* 2013 Oct;23(5):841–6. <doi:10.1016/j.conb.2013.04.001>. Epub 2013 Apr 23. PMID: 23623392; PMCID: PMC4552336.

12 Samson, DR. (2021). 'The human sleep paradox: the unexpected sleeping habits of Homo sapiens.' *Annual Review of Anthropology*, 50(1), 259–274.

13 Kappeler, PM. (1998). 'Nests, tree holes, and the evolution of primate life histories.' *American Journal of Primatology*, 46(1), 7–33. <doi.org/10.1002/(sici)1098-2345(1998)46:1%3C7::aid-ajp3%3E3.0.co;2-#>

14 Ibid. Nunn, CL & Samson, DR. (2018). 'Sleep in a comparative context: Investigating how human sleep differs from sleep in other primates.' *Am J Phys Anthropol.* 2018 Jul;166(3):601–612. <doi:10.1002/ajpa.23427>. Epub 2018 Feb 14. PMID: 29446072.

15 Samson, DR et al. (2017). 'Chronotype variation drives night-time sentinel-like behaviour in hunter–gatherers.' *Proceedings of the Royal Society B: Biological Sciences* 284.1858 (2017): 20170967.

16 Gowlett, JAJ. (2016). 'The discovery of fire by humans: A long and convoluted process.' *Phil. Trans. R. Soc.* B3712015016420150164.

17 Nunn, CL & Samson, DR. (2018). 'Sleep in a comparative context: Investigating how human sleep differs from sleep in other primates.' *American Journal of Physical Anthropology* 166.3 (2018): 601–612.

18 Yamazaki, R et al. (2020). 'Evolutionary origin of distinct NREM and REM sleep.' *Frontiers in Psychology* 11 (2020): 3599.

19 Dresler, M, Koch, SP, Wehrle, R, Spoormaker, VI, Holsboer, F, Steiger, A, ... & Czisch, M. (2011). 'Dreamed movement elicits activation in the sensorimotor cortex'. *Current Biology*, *21*(21), 1833–1837.

20 Flanagan, O. (1995). 'Deconstructing dreams: The spandrels of sleep.' *The Journal of Philosophy* 92.1 (1995): 5–27.

21 Hoel, E. (2021). 'The overfitted brain: Dreams evolved to assist generalization.' *Patterns* 2.5 (2021).

22 Kilius, E et al. (2021). 'Pandemic nightmares: COVID-19 lockdown associated with increased aggression in female university students' dreams.' *Frontiers in Psychology* 12 (2021): 644636.

23 Bottary, R, Seo, J, Daffre, C, Gazecki, S, Moore, KN et al. (2020). 'Fear extinction memory is negatively associated with REM sleep in insomnia disorder.' *Sleep*. 2020 Jul 13;43(7):zsaa007. <doi: 10.1093/sleep/zsaa007>. PMID: 31993652; PMCID: PMC7355402.

24 Hill, TD, Trinh, HN, Wen, M & Hale, L. (2016). 'Perceived neighborhood safety and sleep quality: A global analysis of six countries.' *Sleep Med.* 2016 Feb;18:56–60. <doi:10.1016/j.sleep.2014.12.003>. Epub 2014 Dec 15. PMID: 25616390.

25 Raap, T, Pinxten, R & Eens, M. (2015). 'Light pollution disrupts sleep in free-living animals.' *Scientific Reports* 5.1 (2015): 13557.

Chapter 9 Why do we have inner lives?

1 Nagel, T. (1974). 'What is it like to be a bat?' *The Philosophical Review*. (1974): 435–450.

2 Sattin, D, Magnani, FG, Bartesaghi, L, Caputo, M, Fittipaldo, AV et al. (2021). 'Theoretical models of consciousness: A scoping review.' *Brain Sci.* 2021 Apr 24;11(5):535. <doi:10.3390/brainsci11050535>. PMID: 33923218; PMCID: PMC8146510.

3 World, U. (2018b). 'Chengjiang Fossil Site.' Unesco. <whc.unesco.org/en/list/1388>

4 Dickson, BV, Clack, JA, Smithson, TR et al. (2021). 'Functional adaptive landscapes predict terrestrial capacity at the origin of limbs.' *Nature* 589, 242–245 (2021). <doi.org/10.1038/s41586-020-2974-5>

5 *The State of World Fisheries and Aquaculture 2022: Towards blue transformation*. Rome, FAO. (2022). <doi.org/10.4060/cc0461en>

6 Brown, R, Lau, H, & LeDoux, JE. (2019). Understanding the higher-order approach to consciousness. Trends in cognitive sciences, 23(9), 754–768. <www.sciencedirect.com/science/article/pii/S1364661319301615#bb0015>

7 Crick, FC & Koch, C. (1990). 'Towards a neurobiological theory of consciousness.' *Sem. Neurosci.* 2, 263–275 (1990).

8 Ehret, G & Romand, R. (2022). 'Awareness and consciousness in humans and animals – neural and behavioral correlates in an evolutionary perspective.' *Front Syst Neurosci.* 2022 Jul 14;16:941534. <doi:10.3389/fnsys.2022.941534> PMID: 35910003; PMCID: PMC9331465.

9 Humphrey, N. (2023, September 22). 'How did consciousness evolve? – with Nicholas Humphrey.' Youtube. <www.youtube.com/watch?v=9QWaZp_2I1k>

10 Stacho, M, Herold, C, Rook, N, Wagner, H, Axer, M et al. (2020). 'A cortex-like canonical circuit in the avian forebrain.' *Science.* 2020 Sep 25;369(6511):eabc5534. <doi:10.1126/science.abc5534>. PMID: 32973004.

11 Baars, BJ. (1988). *A Cognitive Theory of Consciousness*. Cambridge: Cambridge Cambridge University Press.

12 Zacks, O & Jablonka, E. (2023). 'The evolutionary origins of the Global Neuronal Workspace in vertebrates.' *Neuroscience of Consciousness*, 2023(1), 2023, niad020. <doi.org/10.1093/nc/niad020>

13 Baars, BJ, Geld, N, & Kozma, R. (2021). 'Global workspace theory (GWT) and prefrontal cortex: Recent developments.' *Frontiers in psychology, 12*, 749868. <www.frontiersin.org/articles/10.3389/fpsyg.2021.749868/full>

14 Salena, MG, Turko, AJ, Singh, A, Pathak, A, Hughes, E et al. (2021). 'Understanding fish cognition: A review and appraisal of current practices.' *Animal Cognition*, 24(3), 395–406. <doi:10.1007/s10071-021-01488-2>

15 Jorge, PE, Almada, F, Gonçalves, AR, Duarte-Coelho, P & Almada, VC. (2012). 'Homing in rocky intertidal fish. Are *Lipophrys pholis* L. able to perform true navigation?' *Anim Cogn*. 2012 Nov;15(6):1173–81. <doi:10.1007/s10071-012-0541-7>. Epub 2012 Aug 5. PMID: 22864924.

16 Salena, MG, Turko, AJ, Singh, A et al. (2021). 'Understanding fish cognition: A review and appraisal of current practices.' *Anim. Cogn* 24, 395–406 (2021). <doi.org/10.1007/s10071-021-01488-2>.

17 Kohda, M et al. (2022). 'Further evidence for the capacity of mirror self-recognition in cleaner fish and the significance of ecologically relevant marks.' *PLoS Biology* 20.2 (2022): e3001529.

18 Bshary, R et al. (2006). 'Interspecific communicative and coordinated hunting between groupers and giant moray eels in the Red Sea.' *PLoS Biology* 4.12 (2006): e431.

19 Abbott, A. (2015). 'Clever fish.' *Nature*, 521(7553), 412.

20 Key B. (2015). 'Fish do not feel pain and its implications for understanding phenomenal consciousness.' *Biology & Philosophy*, 30(2), 149–165. < doi.org/10.1007/s10539-014-9469-4>

21 Arbib, MA. (2017). 'How language evolution reshaped human consciousness: A foundational approach.' *Biophysics of Consciousness: A foundational approach*. Singapore: World Scientific, 87–128.

22 Solms, M & Friston, K. (2018). 'How and why consciousness arises: Some considerations from physics and physiology.' *Journal of Consciousness Studies* 25.5-6 (2018): 202–238.

23 Aleman, B & Merker, B. (2014). 'Consciousness without cortex: A hydranencephaly family survey.' *Acta Paediatrica* 103.10 (2014): 1057–1065.

24 Merker, B. (2007). 'Consciousness without a cerebral cortex: A challenge for neuroscience and medicine.' *Behav Brain Sci.* 2007 Feb;30(1): 63–81; discussion 81–134. <doi:10.1017/S0140525X07000891>. PMID: 17475053.

25 Ibid.

26 Cerqueira, M, Millot, S, Castanheira, MF, Félix, AS, Silva, T et al. (2017). 'Cognitive appraisal of environmental stimuli induces emotion-like states in fish.' *Scientific Reports*, 7(1), 13181.

27 Akinrinade, I, Kareklas, K, Teles, MC, Reis, TK, Gliksberg, M et al. (2023). 'Evolutionarily conserved role of oxytocin in social fear contagion in zebrafish.' *Science*, 379(6638), 1232–1237.

28 Bateson, M, Desire, S, Gartside, SE et al Wright, GA. (2011). 'Agitated honeybees exhibit pessimistic cognitive biases.' *Current Biology*, 21(12), 1070–1073. <doi.org/10.1016/j.cub.2011.05.017>

29 Rose, JD et al. (2014). 'Can fish really feel pain?' *Fish and Fisheries* 15.1 (2014): 97–133.

30 Ibid.

31 Ginsburg, S, Jablonka, E, & Zeligowski, A. (2019). *The Evolution of the Sensitive Soul: Learning and the origins of consciousness*. The Mit Press. Birch, J, Ginsburg, S, & Jablonka, E. (2020). 'Unlimited associative learning and the origins of consciousness: A primer and some predictions.' *Biology & philosophy*, 35, 1–23.

32 McGraw, MB. (1941). 'Neural maturation as exemplified in the changing reactions of the infant to pin prick.' *Child Development*, 1941, vol. 12, no. 1, 31–42. JSTOR, <doi.org/10.2307/1125489>.

33 Rodkey, EN & Pillai Riddell, R. (2013). 'The infancy of infant pain research: The experimental origins of infant pain denial.' *J Pain*. 2013 Apr;14(4):338–50. <doi:10.1016/j.jpain.2012.12.017> PMID: 23548489.

34 Ibid.

35 Anand, KJS & Hickey, PR. (1987). 'Pain and its effects in the human neonate and fetus.' *New England Journal of Medicine*. 317.21 (1987): 1321–1329.

36 Goksan, S et al. (2015). 'fMRI reveals neural activity overlap between adult and infant pain.' *eLife* 4 (2015): e06356.

37 Guillory, JD. (1968). 'The pro-slavery arguments of Dr. Samuel A Cartwright.' *Louisiana History: The Journal of the Louisiana Historical Association*, vol. 9, no. 3, 1968, 209–227. JSTOR, <www.jstor.org/stable/4231017>. Accessed 15 Jan. 2024.

38 Hoffman, KM, Trawalter, S, Axt, JR & Oliver, MN. (2016). 'Racial bias in pain assessment and treatment recommendations, and false beliefs about biological differences between blacks and whites.' *Proc Natl Acad Sci U S A*. 2016 Apr 19;113(16):4296–301. <doi:10.1073/pnas.1516047113>. Epub 2016 Apr 4. PMID: 27044069; PMCID: PMC4843483.

39 Diggles, BK et al. (2024). 'Reasons to be skeptical about sentience and pain in fishes and aquatic invertebrates.' *Reviews in Fisheries Science and Aquaculture* 32.1 (2024): 127–150.

40 Rose, JD et al. (2014). 'Can fish really feel pain?'

41 Suzuki, DG. (2022). 'The anthropic principle for the evolutionary biology of consciousness: Beyond anthropocentrism and anthropomorphism.' *Biosemiotics* 15, 171–186 (2022). <doi.org/10.1007/s12304-022-09474>

42 Nagel, T. (1974). 'What is it like to be a bat?' 435–450.

43 Gutfreund, Y. (2018). 'The mind-evolution problem: the difficulty of fitting consciousness in an evolutionary framework.' *Frontiers in Psychology* 9 (2018): 1537.

44 Birch, J, Schnell, AK, & Clayton, NS. (2020). 'Dimensions of animal consciousness.' Trends in Cognitive Sciences, 24(10), 789–801. <www.cell.com/trends/cognitive-sciences/fulltext/S1364-6613(20)30192-3>

45 Trewavas, A. (2021). 'Awareness and integrated information theory identify plant meristems as sites of conscious activity.' *Protoplasma* 258, 673–679 (2021). <doi.org/10.1007/s00709-021-01633-1>

46 Trewavas, A et al. (2020). 'Consciousness facilitates plant behavior.' *Trends in Plant Science* 25.3 (2020): 216–217.

47 Gagliano, M et al. (2014). 'Experience teaches plants to learn faster and forget slower in environments where it matters.' *Oecologia* 175 (2014): 63–72.

48 Gagliano, M et al. (2016). 'Learning by association in plants.' *Scientific Reports* 6.1 (2016): 38427.

49 Mallatt, J, Blatt, MR., Draguhn, A, Robinson, DG., & Taiz, L. (2021). 'Debunking a myth: plant consciousness.' *Protoplasma*, 258(3), 459–476.

50 Aratani, Y, Uemura, T, Hagihara, T et al. (2023). 'Green leaf volatile sensory calcium transduction in *Arabidopsis*.' *Nat Commun* 14, 6236 (2023). <doi.org/10.1038/s41467-023-41589-9>

Interviews

Author interviews for Why Are We Like This? *unless otherwise specified.*

Introduction
Suosaari, E. (2023). Page 4.
Van Kranendonk, M. (2023). Quotes not included.
Lechte, M. (2023). Quotes not included.

Chapter 1 Why do we care?
Thalmann, S. (2023). Page 13.
Aktipis, A. (2023). Page 31.

Chapter 2 Why do we have sex?
Wilner, A. (2023). Page 45.
Bonduriansky, R (2023).Page 45.
Ambur, O (2023). Page 47.
Logsdon, J (2023). Page 51.
Michod, R (2023). Page 53.
Otto, S (2023). Page 58.
Miller, S (2023). Page 64.

Chapter 3 Why do we have males, females and other sexes?
Kokko, H (2022). Page 69.
Tang-Martinez, Z (2023). Page 75.
Yasui, Z (2023). Interview via email. Page 82.
Hopkins, B (2023). Page 85.
Cameron, E (2019). Page 95.
Cameron, E (2023). Page 95.
Servedio, R (2023). Quotes not included.

Chapter 4 Why do we make love?
Monk, J (2023). Page 112.
Lerch, B (2023). Page 114.
Pavličev, M (2023). Page 120.
Bode, A (2023). Page 133.
Sandel, A (2023). Page 136.
Rodrigues, M (2023). Page 141.

Chapter 5 Why do we get cancer?
Rodrigues, M (2022). Page 146.
Rodrigues, M (2023). Page 146.
Aktipis, A (2023). Page 154.
Nedelcu, A (2023). Page 154.
Caulin, A (2023). Page 154.
Schiffman, J (2023). Page 161.
Abegglen, L (2023). Page 161.
Schroeder, A (2023). Page 163.
Gatenby, R (2022). IPage 166.
Gatenby, R (2023). Page 166.
Freytag, S (2023). Page 166.
Ujvari, B (2022). Quotes not included.
Vendramin, R (2022). Quotes not included.

Chapter 6 Why do we age?
Lemoine, M (2024). Page 191.
Bonduriansky, R (2023). Page 200.
Hoffman, J (2024). Page 203.

Chapter 7 Why do we drink?
Clites, B (2023). Page 220.
Dominy, N (2023). Page 225.
Dunbar, R (2023). Page 236.
Melin, A (2023). Quotes not included.

Chapter 8 Why do we sleep?
Rummer, J (2023). Page 246.
Hasenei, A (2023). Page 246.
Hayashi, Y (2023). Page 252.
Samson, D (2023). Page 258.
Burns, C (2023). Quotes not included.

Acknowledgments

Very few books are the product of the author alone, and this is particularly true of *Why Are We Like This?* Much like the research the book explores, my writing is evidence that collaboration is necessary to create anything with complexity. There are many people to thank for their support, advice, and friendship that kept me inspired and on track during the wonderful challenge of writing my first book.

To Jackie Kerin – thank you for accompanying me on the first research trip for my the book, and for all your reading, feedback, and editorial notes that gave me confidence in the work, and always filling my world with stories. To John Kean – thank you for shaping my curiosity for the world and for getting me into the water from the early 90s.

To Lucy Cutting – thank you for reading my chapters as they were written and wading through many early unpolished versions. The emotional support you provided to me consistently, even as the late-night nerves kicked in, will never be forgotten.

To Tegan Taylor – thank you for teaching me how to write for a general audience, it is a gift I will always cherish. And thanks for reading select chapters and giving me sound advice for this project.

To my test readers who perused chapters and provided their thoughts and feedback, I am so grateful to you all, including Dr Naomi Koh Belic, Lee Constable, Max Nilssen, and Ness Finn. Ness, a special thank you for your ongoing support, from reading my submission to helping me on the day the manuscript was due. Your friendship and counsel were so valuable. And Max thank you for swim, moral and Sadie support on top of the maths wisdom.

A massive thank you to Dr Jenny Phan for joining me on a mission to Townsville, and a good job encouraging me to partake of pina coladas while we wrote in the tropics. You've always pushed me to challenge my assumptions about the world, and think clearly, carefully and scientifically. To Lucy Kervin-McDermott, over 20 years of punning at each other has honed my language skills, so thanks for that, but more seriously thank you for being such an incredible friend. To Veronica Ferrie, since we were little you have always celebrated my scientific obsessions, and I have always felt your belief in me. I don't think this book would have ever existed without that base. My love to you all.

To my Hobart community, thank you for always being there when I needed you. Olivia Tunks, your reliable support and friendship is so deeply appreciated. To my ABC Radio Hobart colleagues, thank you for standing by me, I'm very lucky to have a workspace where creativity, eccentricity, and ideas are celebrated. Also thank you to the team at ABC Science and to Beaker Street Festival. And to Karl K and Mary D, for trusting my scientific chops and encouraging me to reach larger audiences, and a place to stay by the beach.

To my publisher Harriet McInerney, thank you for believing in this project, for your patience as I talked out my ideas and your wise counsel over many ramens. To my editor Linda Funnell, thank you for taking on such a wide-ranging book, tackling so many topics and helping my words to shine.

Additional thanks to Trish Striker, Holly, and Levi Joule.

There were many researchers interviewed for this book whose quotes did not make it to the page, but whose ideas certainly informed the text. Many thanks to Martin Van Kranendonk, Maxwell Lechte, Maria Servedio, Roberto Vendramin (who I interviewed extensively twice!), Beata Ujvari, Amanda Melin, Courtney Burns, and Lila Landowski. Thanks also to the Australasian Evolution Society for so generously welcoming a science journo at your 2022 conference.

Lastly, thank you to my wonderful family. My partner Natalie and Sadie the schnauzer. Thank you for the loving and supportive home we have created together. You are the foundation this book is built on, and I will always appreciate that. You are both the very best.